Clinical Medicine and the Nervous System
Series Editors: John P. Conomy and Michael Swash

Titles in the series already published:

Headache
Richard Peatfield

Epilepsy: Electroclinical Syndromes
Edited by Hans Lüders and Ronald P. Lesser

The Heart and Stroke
Edited by Anthony J. Furlan

Forthcoming titles in the series:

Imaging of the Nervous System
Edited by Paul Butler

Dizziness and Disequilibrium
Thomas Brandt

Electrophysiological Diagnosis
E. M. Sedgwick

Neurological Emergencies
John P. Conomy, Martin S. Schwartz and Michael Swash

Malignant Brain Tumours
Edited by David G. T. Thomas and D. I. Graham

Hydrocephalus and the Cerebrospinal Fluid
Edited by J. D. Pickard and R. O. Weller

Hierarchies in Neurology

A Reappraisal of a Jacksonian Concept

Edited by
Christopher Kennard and Michael Swash

With 38 Figures

Springer-Verlag
London Berlin Heidelberg New York
Paris Tokyo

Christopher Kennard, BSc, MBBS, PhD, FRCP
Consultant Neurologist, Neurology Department, The London Hospital, Whitechapel, London E1 1BB, UK.

Michael Swash, MD, FRCP, MRCPath
Consultant Neurologist, Neurology Department, The London Hospital, Whitechapel, London E1 1BB, UK.

Series Editors

John P. Conomy, MD
Chairman, Department of Neurology, The Cleveland Clinic Foundation, 9500 Euclid Avenue, Cleveland, Ohio 44106, USA.

Michael Swash, MD, FRCP, MRCPath
Consultant Neurologist, Neurology Department, The London Hospital, Whitechapel, London E1 1BB, UK.

ISBN-13:978-1-4471-3149-6 e-ISBN-13:978-1-4471-3147-2
DOI: 10.1007/978-1-4471-3147-2

British Library Cataloguing in Publication Data
Kennard, Christopher
Hierarchies in neurology.
1. Medicine. Neurology
I. Title II. Swash, Michael, *1939–*
616.8
ISBN-13:978-1-4471-3149-6

Library of Congress Cataloging-in-Publication Data
Hierarchies in neurology: a reappraisal of a Jacksonian concept
Christopher Kennard, Michael Swash (eds.)
p. cm. Includes bibliographies and index
ISBN-13:978-1-4471-3149-6
1. Neurology—Philosophy. 2. Brain—Pathophysiology—Philosophy. 3. Jackson, J. Hughlings (John Hughlings), 1835–1911.
I. Kennard, Christopher. II. Swash, Michael.
[DNLM: 1. Jackson, J. Hughlings (John Hughlings), 1835–1911. 2. Neurology.
WL 100 H6335] RC346.H54 1989 616.8'01—dc19 DNLM/DLC
for Library of Congress 88–39153

Softcover reprint of the hardcover 1st edition 1989

Filmset by Wilmaset, Birkenhead, Wirral

2128/3916–543210 Printed on acid-free paper

The London Hospital in 1870 drawn from the west, from across the Whitechapel Road, showing the new Grocer's wing (still open) on the occasion of its opening.

Series Editors' Foreword

Hughlings Jackson's ideas were of fundamental importance in the establishment of neurology as a clinical discipline, but he occupies a somewhat remote and mysterious place in the education and everyday experiences of modern neurologists and neuroscientists. Neurology evolved from the work of many scientists in many different countries, as can be attested by a glance at the reference lists cited by Gowers, Dejerine, Oppenheimer and Kinnier Wilson in their monumental textbooks. The neurological literature may be growing, but an enormous amount of it is simply no longer read, or simply ignored. The cynical journal editor can sometimes be heard to cite "the ten year rule", by which re-publication of certain observations is expected every ten years simply because a new generation of clinician-investigators has not read the older literature.

Jackson's contributions do not fall into this category. Here was a man who observed carefully, thought hard about the significance and implications of his observations, compared them with those of others, and related them to the pervading philosophical and biological theories of his day. Many of his ideas still form the basis of our understanding of common clinical phenomena, especially in the field of epilepsy. We live in an era in which clinical neurology and neuroscience are again coming close together, as it becomes possible to carry out experiments safely and relatively non-invasively in the living human subject, using PET scanning, MRI and the newer neurophysiological methods for exploration of the nervous system. In Jackson's day clinical neurology was itself the cutting edge of neuroscience, and we again are beginning to experience the excitement of the immediate inter-related relevance of neuroscience and neurology.

In this book some of the issues raised by Jackson are reviewed both in relation to his own contributions, and in their wider and more immediate relevance to modern neurology. A careful study of the history of neurological concepts is salutary and revealing to those concerned with current problems; there is much to be learnt from our predecessors. This volume, therefore, is welcomed as part of the *Clinical Medicine and the Nervous System* series, as providing unusual insights into neurology. We trust it will provide instruction and pleasure.

Cleveland, Ohio and London
1988

John P. Conomy
Michael Swash

Preface

The sesquicentennial of Hughlings Jackson's birth coincided with a combined meeting of the Association of British Neurologists and the American Neurological Association at The London Hospital. Hughlings Jackson made his clinical observations on the wards of The London Hospital and at the National Hospital for Nervous Diseases. Queen Square, and this anniversary seemed an appropriate opportunity to re-evaluate his contributions to neurology and to neuroscience. Even a brief survey of Jackson's publications reveals the astonishing range of his observations, but an underlying thread can be seen in his approach to the problem of the functional organisation of the brain. The hierarchical concept, derived from the new and fashionable evolutionary ideas that swept through scientific thought during the latter part of the nineteenth century was particularly important in determining Jackson's approach to neurology. Indeed, so persuasive were his ideas that they were rapidly assimilated into the developing subject, so that, in many respects, Jackson can be said to be the father of British (or Anglo-American) neurology.

This book has itself evolved from that conference, which provided the opportunity for scientists and clinicians to meet and to interact in their views on the nervous system. The views expressed in this book have themselves evolved since the time of that meeting and represent attempts to reconcile the hierarchical concept of brain function still embodied in the thinking of many clinical neurologists with the exciting new concepts of distributive control systems, involving close sensorimotor interactions at multiple anatomical levels in the central nervous system, implicit in more recent contributions derived from the non-clinical neurosciences. The latter are beginning to exert a strong influence on clinical practice.

This reappraisal of Jacksonian concepts of the brain therefore has appeal to clinicians and neuroscientists alike. The book is organised in five sections, representing historical aspects, contributions on consciousness and memory, epilepsy, sensory systems and the motor system respectively. After some consideration we decided it was inappropriate to try to re-evaluate Jackson's views on language in the brief format available in this volume. The book therefore contains a unique evaluation of the hierarchical concept of brain function in relation to the major problems of contemporary behavioural neuroscience; this should be of interest to students of both clinical neurology and of the non-clinical neurosciences. We hope that it will serve to bring together these two branches of science.

London
1988

Christopher Kennard
Michael Swash

Contents

Contributors

A. Baddeley, PhD
MRC Applied Psychology Unit, 15 Chaucer Rd, Cambridge CB2 2EF, UK

Macdonald Critchley, CBE, MD, FRCP, (Hon) FACP
Hughlings House, 12 Mill Lane, Nether Stowey, Bridgwater, Somerset TA5 1NL, UK

J. C. Eccles, MB.BS, MA, DPhil, FRS
CH 6611 Contra (TI) Switzerland

R. A. Henson, MD, FRCP
The Nab, Church Rd, Newnham-on-Severn, Gloucestershire

R. J. Joynt, MD, PhD
Department of Neurology, School of Medicine and Dentistry, Medical Center, The University of Rochester, 601 Elmwood Avenue, Rochester, NY 14642, USA

C. Kennard, BSc, MB.BS, PhD, FRCP
Neurology Department, The London Hospital, Whitechapel, London E1 1BB, UK

P. B. C. Matthews, MD, DSc, FRS
University of Oxford, Laboratory of Physiology, Parks Road, Oxford OX1 3PT, UK

B. S. Meldrum, MB.BChir, PhD
Department of Neurology, Institute of Psychiatry, De Crespigny Park, Denmark Hill, London SE5 8AF, UK

E. H. Reynolds, MD, FRCP, FRCPsych
Department of Neurology, Maudsley and King's College Hospitals, Denmark Hill, London SE5 9RS, UK

E. T. Rolls, MA, DPhil, DSc
University of Oxford, Department of Experimental Psychology, South Parks Road, Oxford OX1 3UD, UK

J. F. Stein, MA, MSc, BM.BChir, MRCP
University of Oxford, University Laboratory of Physiology, South Parks Road, Oxford OX1 3PT, UK

M. Swash, MD, FRCP, MRCPath
Neurology Department, The London Hospital, Whitechapel, London E1 1BB, UK

M. R. Trimble, FRCP, FRCPsych
Department of Psychiatry, The National Hospitals, Queen Square, London WC1 3BG, UK

Å. B. Vallbo, MD, PhD
Department of Physiology, University of Göteborg, S-400 33, Göteborg, Sweden

Priv. Doz, Dr. med. Claus-W. Wallesch,
Abteilung Klinische Neurologie und Neurophysiologie, Universität Freiburg, Hansastr. 9, D-7800 Freiburg, Federal Republic of Germany

A. Wilkins, DPhil
MRC Applied Psychology Unit, 15 Chaucer Road, Cambridge CB2 2EF, UK

Section I

HISTORICAL

Chapter 1

John Hughlings Jackson: A Historical Introduction

M. Swash

John Hughlings Jackson (Figs. 1.1, 1.2) was born 150 years ago on 4 April 1835 at Green Hammerton in Yorkshire. He died on 7 October 1911 and was buried in Highgate Cemetery. Jackson's contributions to neurology (Taylor 1931/32) were extraordinarily wide-ranging and had far-reaching influences on his contemporaries (Broadbent 1903; Ferrier 1912; Buzzard 1934; Harris 1935), and on those who followed him. Despite this formative role in the establishment of clinical neurology as a specialty in its own right, his contributions in a broader sense to studies of the central nervous system were far less widely recognised in his own lifetime. Indeed, he received none of the honours, for example a knighthood, that were bestowed on so many of his less distinguished contemporaries. The reasons for this are not hard to find in his peculiarly private personality and, perhaps, in the convoluted style of his prose. Further, he seems not to have been interested in developing an extensive and fashionable private practice and consequently was less well known in London society than were many of his contemporaries.

Education

He was educated at schools in Green Hammerton (Broadbent 1903; Dewhurst 1982), and at Nailsworth in Gloucestershire until about the year 1850, when he was apprenticed to a Dr William Charles Anderson in a practice in Stonegate, York. Dr Anderson lectured in the York Medical School and Jackson spent about 5 years in York before coming to London to spend a short time studying at St Bartholomew's Hospital with Sir James Paget. In 1856 he qualified as a Licenciate of the Society of Apothecaries and a Member of the Royal College of Surgeons and then worked for 3 years, until 1859, as House Physician at the York Dispensary. There he was introduced to the study of mental diseases by Daniel Tuke [1827–1895] and Thomas Laycock [1812–1876]. Tuke was physician to The

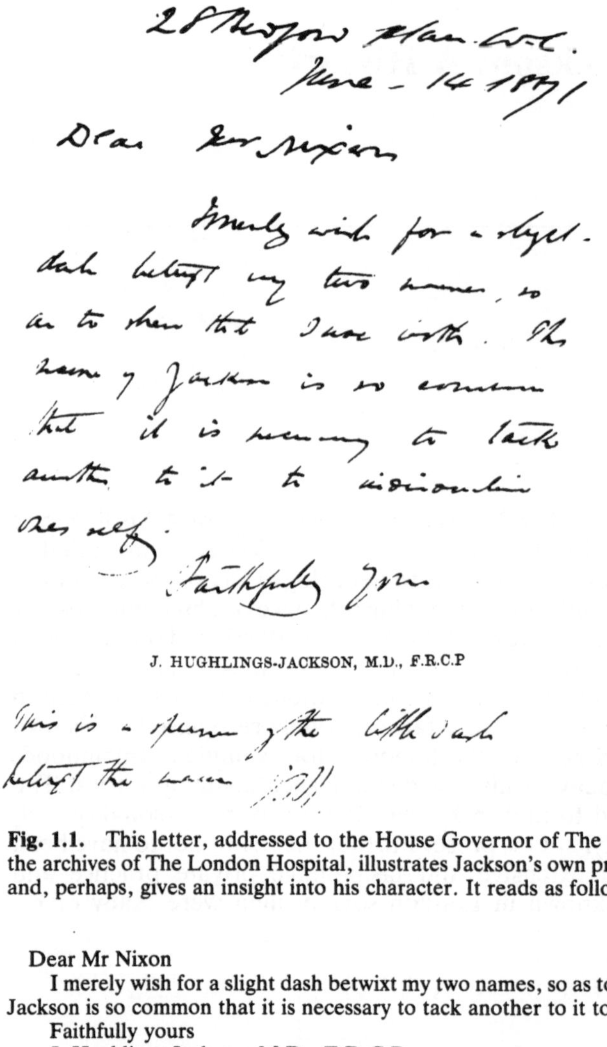

28 Bedford Place W.C.
June - 14 1871

Dear Mr Nixon

I merely wish for a slight dash betwixt my two names, so as to show that I use both. The name of Jackson is so common that it is necessary to tack another to it to individualise ones self.

Faithfully yours

J. HUGHLINGS-JACKSON, M.D., F.R.C.P

This is a specimen of the little dash betwixt the names JH-J.

Fig. 1.1. This letter, addressed to the House Governor of The London Hospital, and preserved in the archives of The London Hospital, illustrates Jackson's own preference in the spelling of his name and, perhaps, gives an insight into his character. It reads as follows:

28 Bedford Place. W.C.
June 14 1871

Dear Mr Nixon
I merely wish for a slight dash betwixt my two names, so as to show that I use both. The name of Jackson is so common that it is necessary to tack another to it to individualise one's self.
Faithfully yours
J. Hughlings Jackson, M.D., F.R.C.P.
This is a specimen of the little dash
betwixt the names
JH-J.

Retreat, a hospital for mental diseases that had been founded by his great grandfather. Tuke had been educated at St Bartholomew's Hospital in London and at the University of Heidelberg, and Laycock, after attending University College, London, spent 2 years studying neuropsychiatric disorders in Paris and a year at Göttingen. Laycock, in particular, had been much influenced by Marshall Hall and the doctrine of reflex action and was clearly an important influence, in his turn, on Jackson. In 1859 Jackson was introduced by Samuel North, a surgeon in York, to another former student of the York Medical School, Jonathan Hutchinson, who had been appointed Assistant Surgeon to The London Hospital

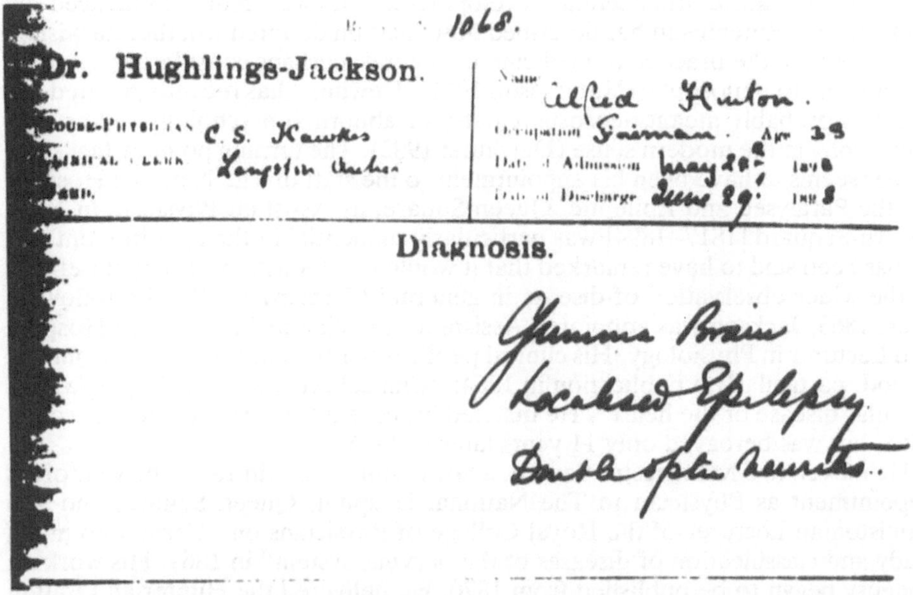
1068.

Dr. Hughlings-Jackson.

House-Physician C. S. Hawkes

Name Alfred Hunton.

Occupation Fireman. Age 33

Date of Admission May 28 1898

Date of Discharge June 29 1898

Diagnosis.

Gumma Brain

Localised Epilepsy

Double optic neuritis.

Fig. 1.2. An example of the front sheet of the case notes of a patient admitted to The London Hospital under Dr Hughlings Jackson's care. Some of these notes have been preserved on microfilm. In accordance with the practice of the time a wide variety of patients came under his care, many not suffering from neurological disease.

in that year (Hutchinson 1911). Hutchinson went on to pursue an influential and successful career at The London Hospital (Clark-Kennedy 1962), becoming Surgeon to the Hospital in 1863, Fellow of the Royal Society in 1881 and President of the Royal College of Surgeons in 1889. For 3 years Jackson lived with Jonathan Hutchinson at the latter's house at 14 Finsbury Circus.

Life in London

On Jackson's arrival in London in 1859, Hutchinson arranged for him to be appointed a Lecturer in Pathology, Morbid Anatomy and Histology at The London Hospital Medical School. In the same year Hutchinson also helped Jackson obtain attachments as Assistant Physician to the Metropolitan Free Hospital, and to the Islington Dispensary. Jackson and Hutchinson together contributed medical reports to the *Medical Times and Gazette* during this period. In 1860 Jackson became clinical assistant to Mr Alfred Poland, at Moorfield's Eye Hospital and in the same year obtained the MD of St Andrew's University. In 1861 he became a Member of the Royal College of Physicians and moved to a house near Hutchinson's in Finsbury Circus (Clark-Kennedy 1962; Dewhurst 1982).

During this 3-year period from 1859 to 1861, while Jackson was feeling his way towards his future intellectual developments, he was much influenced by Hutchinson. Hutchinson has described how Jackson doubted whether he wished to continue in the practice of medicine in general, perhaps considering devoting his intellect to philosophy (Hutchinson 1911). Dewhurst has recently pointed out that this probably meant neurospsychiatry or abnormal psychology rather than philosophy in the modern sense (Dewhurst 1982). The turning point in Jackson's career seems to have been his appointment to the staff of The National Hospital for the Paralysed and Epileptic, Queen Square, as Assistant Physician in 1862. Brown-Séquard [1817–1894] was particularly influential in the appointment and he has been said to have remarked that it would be "foolish to waste your efforts in the wider observation of disease in general" (Buzzard 1934). The following year, 1863, Jackson was appointed Assistant Physician to The London Hospital and Lecturer in Physiology. His clinical publications in neurology begin from this period, particularly a publication in 1864: "Clinical remarks on hemiplegia with valvular disease of the heart". He married his cousin Elizabeth Dade Jackson in 1865, but was bereaved only 11 years later in 1876.

He moved to 3 Manchester Square, a house still extant, in 1867, the year of his appointment as Physician to The National Hospital, Queen Square, and was Goulstonian Lecturer of the Royal College of Physicians on "Certain points of study and classification of diseases of the nervous system" in 1869. His work on epilepsy began to be published from 1870. He delivered the Hunterian Oration on "Physiological aspects of education" in 1872 and he was appointed Physician to The London Hospital in 1874. His enthusiasm for the ophthalmoscope was communicated in his annual oration to the Medical Society of London in 1877 and he was elected Fellow of the Royal Society in 1878, the year of the foundation of *Brain*. He was one of the three founder editors of that journal. He continued to develop his ideas on epilepsy and became increasingly interested in applying evolutionary concepts, as put forward by Herbert Spencer, to the classification and understanding of neurological diseases from about 1880. His Croonian lectures, given in 1884, "On evolution and dissolution of the nervous system" are a particular landmark in this aspect of his ideas.

He retired from the staff of The London Hospital in 1894. The presentation on this occasion was made by Sir James Paget, who presented Jackson with the portrait by Lance Calkin that now hangs in The London Hospital Medical College. A copy of this portrait hangs on the walls of The National Hospital, Queen Square. He retired from active clinical practice in 1906, the date of his retirement from The National Hospital, Queen Square.

The peculiarities of Jackson's personality in the years after the death of his wife are particularly clearly described by Sir James Taylor (Taylor 1931/32) and, in a Schorstein Lecture delivered at The London Hospital, by Sir E. Farquhar Buzzard (Buzzard 1934):

> I remember him in those days as the generous, kind-hearted but rather grave family friend or pseudo-uncle whose mind seemed to be in a state of constant conflict between his desire to give pleasure and his fear or being bored or bound. I do not think it can be easy for anyone who was not brought intimately in contact with him to understand the feelings of reverent affection which all of the younger generation who had enjoyed that good fortune always entertained for the great personality of Hughlings Jackson. He was so simple, yet so stimulating, and extraordinarily careful in his words so that he might convey his exact meaning. No doubt this is the reason why his papers are so often overloaded with footnotes. I once heard him say with a smile that he was distressed to think that in speaking of a man as being covered with a rash from head to foot he had been speaking unscientifically, for what he should have said was, "From the tip of his nose to his perineum"!

Buzzard concluded that "Jackson's pre-eminence was due to the fact that not only had he no superior in the work of making detailed and accurate observations, but no real rival in the art of generalisation or, in other words, of integrating those observations so as to produce an harmonious design of the nervous system as a whole".

Jackson had a peculiarly intense relationship with his junior colleagues, well exemplified in the Memoirs written by them after his death (Ferrier 1912; Mercier 1912; Editorial 1915; Head 1915; Buzzard 1934; Harris 1935). According to Buzzard (1934), Jackson's visits for ward rounds, both at The London Hospital, and at The National Hospital, "were quite unfixed in regard to time", unlike those of his colleagues. Buzzard elaborated his description:

> They were frequent but irregular in their frequency, depending on his interest in some particular patient or problem at the moment. The ward round was neither formal nor prolonged and it was impossible to predict to which case he would devote most attention. If one was foolish enough to remark that "This is just an ordinary case of hemiplegia" he might spend the next twenty minutes in demonstrating its unusual features or in explaining how such features illustrated one or other of his theories in regard to the evolution and dissolution of the nervous system. Quite suddenly his attention and interest would seem to fail; he would escape from the ward, but before we parted at the front door of the hospital he would say quietly, "I will find you a reprint setting out my views on the subject we have been discussing, but you need not believe them." He had a great belief in the value of sitting quietly at the end of the day, notebook in hand, to allow unconscious reasoning to form new associations in his mind and to mull over the intellectual implications of his clinical experiences.

Jackson and Spencer

In order to understand the origin of Jackson's concepts (Riese 1956) the pivotal role of Herbert Spencer must be considered (Riese 1950). Spencer, now little regarded, was the greatest popular philosopher of his day (MacPherson 1901). The period of Jackson's early maturity, from about 1850 onwards, coincides with a period of enormous change in social, religious and scientific attitudes. During this period the eighteenth century metaphysical ideas of Hume and John Stuart Mill gave way to new ideas, in the context of the collapse of the Enlightenment, and the confusion and disarray associated with the upheavals of the French Revolution and the Napoleonic wars that followed. This was the period in which the ideas of Kant and Hegel arose in Germany. In Britain the conflict of science and theology was a factor in philosophy long before Darwin's publication of *The Origin of Species* in 1859. Whewell attempted a universal approach to understanding science and theology in his *History and Philosophy of the Inductive Sciences*, a work that Jackson would certainly have read. J. S. Mill's concept of the Regularity of the Cosmos, with its rules and inductions, was discussed, but nature was still viewed as a machine subject to the rules of the cosmos, in which unity was present only in the supernatural being. Huxley and Bentham both also struggled with these concepts.

Herbert Spencer (Macpherson 1901) was born in 1820 in Derby. He was educated by his father, a schoolteacher, and later by his uncle, the Curate of Hinton, near Bath. However, his studies did not prosper and he decided not to attempt entry to university. In 1837 he took employment as a civil engineer on the London and Birmingham Railway and remained working in this capacity, publishing a number of original papers in the *Civil Engineering Journal*, for 8 years until a slump in the railway industry occurred as the result of various railway disasters. He returned to Derby, to his family home, and then in 1848 became Assistant Editor of *The Economist*, a position he held until 1853. During this time he began to publish his own philosophical ideas. His first pamphlet, *The Proper Sphere of Government*, was published in 1842, and he began to formulate ideas in opposition to the current concepts of invariable laws of social behaviour. He speculated on man's relation to society and on theories of population, based on those of Bentham. During these years he met Mary Ann Evans, later well-known as George Eliot, who was then Assistant Editor of *The Westminster Review*, a journal to which he began to contribute extensively. His *Principles of Psychology* was published in 1855, the *Principles of Sociology* and the *Principles of Biology* thereafter. Most of his books were published at his own expense and it was not until 1874, 24 years after his first publication, that his books began to earn him income. During this time he established a national and international reputation as a philosopher and had a particularly important influence in America, for example on William James, and on the continent of Europe, but far less influence in Britain, possibly because of his lack of any university connection. He was himself much influenced by von Baer, whose law of an ascending organisation in the cosmos from a state of homogeneity to that of heterogeneity was adapted by Spencer to imply a hierarchical organisation within organisms, societies and structures from the most simple, in which all parts are of equal type, to the most complex, in which all parts are of different type but inter-relate to form the organism as a whole. In this sense his evolutionary ideas antedated those of Darwin; however, it is clear that the general concept of evolution was arising in many different ways at this time.

It is not known when Jackson met Spencer but they were certainly acquainted, since Jackson introduced Weir Mitchell to Herbert Spencer in 1898, and some correspondence between them has survived (Critchley 1986). Jackson himself wrote (MacPherson 1901) that he "found Mr. Spencer's Principles of Psychology more useful than any other works of psychology in the study of those diseases of the nervous system which have a mental side. I believe that Mr. Spencer's doctrines of evolution and dissolution are of very great value in the methodical analysis of cases of insanity, and further that, on the bases these doctrines supply, relations of different kinds of disease of the highest cerebral centres to one another can be traced, and also relations of disease of these centres to disease of lower centres of the nervous system." Mercier (1912), formerly one of Jackson's House Physicians at The London Hospital, was also much impressed by Spencer: "My idea of the value of Spencer's work is that he has done for co-ordinations in time what Newton did for co-ordinations in space . . . he has reduced chaos to order. He has at any rate discovered the fundamental principles of these sciences."

Spencer himself summarised some of his underlying ideas in a short passage (MacPherson 1901) that illustrates how important it was to Victorian thinkers in the middle of the nineteenth century to develop a theory that transfigured both science and religion in one philosophy:

> Then it was that there suddenly arose in me the conception that the law which I had separately recognised in various groups of phenomena was a universal law applying to the whole Cosmos; the many small inductions were merged in the large inductions. Only then did I see that the universal cause for the universal transformations was the multiplication of effects, and that they might be deduced from the law of the multiplication of effects.

Jackson's Hierarchical Concepts of Nervous Function

Evolution of function within the nervous system was conceived by Jackson as a passage from the most to the least organised, from the most simple to the most complex and from the most automatic to the most voluntary (Taylor 1931/32). Dissolution was understood to be the reverse of the process of evolution. From the lowest to the highest centres there was increasing complexity (*differentiation*), increasing definiteness (*specialisation*), and increasing *integration* (of function). In general, the higher the centres the more numerous the interconnections of their units (*cooperation*). Jackson's concept of a unit was complex. The concept was that of a unit of constitution, meaning a unit containing structures that implied functions in such a balance of relations that they served harmoniously in complete actions. Jackson believed that this unit of constitution of actions was the same at all levels of the nervous system, thus implying that "the whole of the nervous system and its parts are developed on the same fundamental plan". Thus in a lesion of the nervous system recovery simply represented the continuation of the activity of parts spared within the lesion. Compensation did not mean that nervous tissues took on functions that they had never had before, but that lower levels continued to function in a similar way. A hierarchy of functional strategies was thus revealed.

In the last quarter of the twentieth century these ideas appear antiquated and even archaic; however, they enabled Jackson to formulate a hierarchical concept of function, and of disorder of function, within the nervous system that has been of fundamental importance in the development of ideas since that time. In 1880, James Jackson Putnam, of Boston, accompanied Hughlings Jackson to his lectures at The London Hospital. He gave the following description of their impact:

> It was Dr. Jackson's custom to draw a pyramid upon the blackboard which should stand for the hierarchy of the cerebral functions, the more fundamental of them being represented by the basal portions of the pyramid, the more complex and recently acquired by the apex portion. His idea then was that when the hierarchy of functions represented by this pyramid suffers derangement at any part . . . the attempt at a re-establishment of some sort of equilibrium is always such that the new arrangement tends to safeguard itself by accentuating the more fundamental of its powers, while sacrificing, so far as necessary, the more elaborate.

Jackson's formulation of the functions of the brain, and his distinction between psychological concepts and anatomical data, provided a foundation for clinical diagnosis that continues today. More recently his ideas have been taken up in physiological and anatomical investigations of the brain. These more recent approaches to Jackson's concepts of the brain and its functions were reviewed in a symposium entitled Hierarchies in the Brain, held at The London Hospital on 30–31 October 1985, to celebrate the sesquicentenary of his birth. This book has arisen from some of those discussions.

References

Broadbent W (1903) Hughlings Jackson as a pioneer in nervous pathology. Brain 26:305–366
Buzzard EF (1934) Hughlings Jackson and his influence on neurology. Lancet II:909–913

Clark-Kennedy AE (1962) The London: a study of the voluntary hospital system, vols 1 and 2. Pitman Medical, London
Critchley M (1986) Hughlings Jackson; the man and his time. Arch Neurol 43:435–437
Dewhurst K (1982) Hughlings Jackson and neurology. Sandford, Oxford
Editorial (1915) Hughlings Jackson. Br Med J I:757
Ferrier D (1912) John Hughlings Jackson, 1835–1911. Proc R Soc Med 84:xviii–xxv
Harris W (1985) John Hughlings Jackson, 1835–1911. Postgrad Med J 11:131–134
Head H (1915) Hughlings Jackson on aphasia and kindred affections of speech. Brain 38:1–27
Hutchinson J (1911) The late Dr Hughlings Jackson: a recollection of a life-long friendship. Br Med J II:1551–1554
MacPherson H (1901) Herbert Spencer, the man and his work. Chapman and Hall, London
Mercier C (1912) The late Dr Hughlings-Jackson. Br Med J I:85–86
Riese W (1950) Principles of neurology in the light of history and their present use. Nervous and Mental Disease Monographs, New York
Riese W (1956) The sources of Jacksonian neurology. J Nerv Ment Dis 50:125–134
Taylor J (ed) (1931/32) Selected writings of John Hughlings Jackson, vols 1 and 2. Hodder and Stoughton, London. Reprinted (1958) Basic Books, New York

Chapter 2

Hughlings Jackson: The Man and His Time

Macdonald Critchley

John Hughlings Jackson was a product of the Victorian era, even though the sum and substance of his thinking was never realised in all its splendour during his lifetime. Perhaps he was bringing answers to questions that had not yet been asked. Accordingly, it would be better to consider the ideas that were circulating around him before embarking upon a consideration of the man himself.

In 1809 Lamarck had brought out his *Philosophie Zoologique*, which postulated that in biological structure and function progressive change is due to environmental factors. Thus arose the doctrine of "transformation", which shattered all long-accepted ideas as to the existence of a vital force. In 1827 the work of the embryologist von Baer in Königsberg showed that development proceeds from the homogeneous to the heterogeneous, from simple to complex.

Within this exhilarating climate came, in rapid succession, the contributions of Gall, Charles Lyell, Richard Owen, Thomas Huxley, Russell Wallace, and particularly Darwin, whose great work *On the Origin of Species* [1859] was sold out the day it was published.

Little wonder that G. M. Young has commented that of all the decades in our history a wise man would choose the 1850s in which to be born.

On the sidelines, as it were, stood that remarkable figure Herbert Spencer. Unorthodox in schooling, and vacillant in his choice of a vocation, he early revealed his originality in many ingenious technological inventions. These were the expression of a mind that was bursting with unconventional ideas, while recognising no authority and criticising anything that seemed to be established. This rebel, this iconoclast, was destined to survey the panorama of science and to detect a crucial underlying principle, namely evolution. This was something that permeated not only the discipline of biology, but also sociology, ethics, psychology and philosophy. As Professor Acton has said, in this attempt to synthesise the sciences, Spencer showed a sublime audacity (Acton 1985).

These great men were but some of the innovators who stimulated and inspired young Hughlings Jackson. In the year 1850 he left the Nailsworth Boys' Academy to embark upon a career in Medicine. Over the next few years, he was encouraged by that farsighted physician, Thomas Laycock, one of the first to recognise the importance of Marshall Hall's unpopular views about the reflex activity of the nervous system. When, in 1859, Jackson left his native Yorkshire to settle in London, he lodged in the house of Jonathan Hutchinson, whom he had met at the local hospital in York. Hutchinson was a serious, rigid but kindly man, 7 years Jackson's senior, and it was he who dissuaded Jackson from giving up medicine for academic philosophy.

Another contact that proved to be important was Brown-Séquard, who brought to Jackson's notice an impending vacancy on the staff of The National Hospital for the Paralysed and Epileptic, which had been opened 2 years previously in Queen Square. Jackson was interested, he applied and was elected. So began his connection with that hospital which was to continue from 1862 until 1910, the year before he died.

I do not believe that Brown-Séquard was anything more than a catalyst in Jackson's career. It is unlikely that he greatly influenced Jackson's thinking, as Dr. Greenblatt has asserted (Greenblatt 1965), or that he channelled Jackson's interest into the domain of language. Brown-Séquard was a restless oddity, who resigned from The National Hospital in 1863, and returned to America temporarily.

The single greatest influence was surely Herbert Spencer, whose concept of evolution and dissolution (Duncan 1908) was realised by Jackson as being applicable to the nervous system and its disorders. In some respects, Jackson and Spencer were alike. Both were solitary figures of great intellectual stature. Both were poorly educated, artistically unblest, and of dubious literacy. At this point the resemblance halts.

I do not know whether Jackson and Spencer ever met. Certainly they corresponded and did so for nearly 40 years. However, they were not on terms of friendship, even though they had a common contact in the person of Doctor Bastian. I have by me at home a couple of unpublished letters from Spencer to Jackson. One was dated 26 November 1866. Jackson would have been 31 at that time, and Spencer almost 46. The letter reads as follows:

> My dear Sir,
> I am greatly obliged to you for your letter of Saturday containing the series of interesting facts which you have been at so much trouble in setting down for me; and also for a copy of the Medical Journal containing the remarks to which you draw my attention.
>
> The facts respecting incoherence as you have grouped them appear very suggestive, and would I think, of themselves, when further accumulated and classified, go far towards showing the way in which associated ideas become organised. They seem to me quite in harmony with certain more familiar phenomena that occur in old people whose brains are beginning to fail from (as I presume) enfeebled circulation; as also to certain others presented by brains naturally stupid or temporarily stupified. I mean they are analogous in the sense that defect of cerebral power shows itself in the use of the more *general* symbols in place of the more *special* ones. It seems to me worth consideration whether in some of the cases you name the defect is not more due to the absence of an adequate supply of blood to the particular nervous structure that fails in its action, rather than to a lesion of a nervous structure itself. Any abnormal states of the vasomotor nerves controlling the blood vessels that supply part of the brain must, if it is of such a kind as to greatly diminish the circulation in that part, involve a defect of function; this defect, if not extreme, may show itself in the reversion of that function to a simpler form – a form in which the more general relations of thought which are the more deeply organised, alone remain possible.
>
> This view seems to be quite in harmony with your criticism in heredity transmission, which so far as I can judge from partial perusal, seem to me very well grounded. I propose to reserve them in the hope of hereafter turning them to use.
>
> Very truly yours,
> Herbert Spencer

This letter suggests that Jackson had supplied Spencer with observations about the dissolution of speech. The article referred to was probably "Notes on the psychology and pathology of language", which appeared in the *Medical Times and Gazette* 1866, p. 659, and which was re-printed in *Brain*, vol. 38.

The second letter was written 37 years later, on 26 February 1903. Spencer was now lonely, frail and depressed, enduring an invalid existence in Brighton. Jackson, too, was a recluse, aged 68, but in fair health. Spencer was almost 83, and was to die some 8 months later.

> Dear Doctor Jackson,
>
> Thank you for your letter giving me the interesting information concerning Weismann's experiments, and the counter-experiments proposed to be carried out.
>
> Now that I am nearly 83, it is not probable that I shall live to see the results, should a result of a noteworthy kind be forthcoming. But I may say that for myself I have never had any belief in the transmission of effects of mutations. My position has ever been that in pursuance of the loss of inheritance as I understand it, there can be transmission only of those changes that are functionally produced – changes produced by increase of function or changes produced by decrease of function. Should it turn out to be otherwise, I shall be greatly surprised. Of course in any case I shall be interested to learn the results should any be realised while I still live.

That letter was dictated, but he signed it, "Herbert Spencer".

There may well be many other letters which passed between them, but in Spencer's massive autobiography there is but scant reference to Jackson. One interesting note, however, is worth quoting, for it shows that Spencer himself profited from some of Jackson's ideas. Writing on 9 January 1883 to Professor Livingston Youmans, his friend in New York, Spencer wrote:

> I enclose some pages from the Medical Times & Gazette sent to me the other day by Doctor Hughlings Jackson. The initiative he made years ago by applying the doctrine of dissolution to interpretation of nervous disorders – an initiative that is now being followed and in that direction – seems likely to lead to other results. The paper is very clearly and conclusively argued; and it is to one just as much a revelation as that which Hughlings Jackson made of the doctrine . . .

So ends that letter.

Of one thing I do feel confident; had Jackson ever met or been in more intimate personal contact with Herbert Spencer, he would not have liked him. Spencer was an arrogant, captious, pedantic hypochondriac. Jackson was modest, kindly and warm hearted. Where Jackson was simple, Spencer was complicated, feared and respected. Jackson too was respected, but loved. Uncomprehended perhaps, partly because of his difficulty with communication. Jackson had a genial sense of humour; Spencer had less than none. And yet they shared an interest. Spencer wrote an essay on the physiology of laughter, and Jackson lectured to the Medical Society of London on the psychology of joking.

Another person who for many years influenced Jackson was Jonathan Hutchinson, Surgeon to The London Hospital. His impact was of a supportive nature. As I have mentioned already, they had first met at the York Medical School where Hutchinson, 7 years older, was the House Surgeon. When Jackson settled in London, he lodged for 3 years with Hutchinson in his house in Finsbury Circus. For a time they were both on the staff of the *Medical Times and Gazette*, and also the new Sydenham Society. Soon they became life-long cronies – I use that term advisedly – and when Jackson married Elizabeth Dade Jackson, his cousin, it was Hutchinson who gave away the bride. Often they would take holidays together or spend weekends exploring the countryside of Surrey, using the opportunity perhaps for revising papers or preparing lectures. Although Jackson did not share his friend's pietism, or his love of poetry, they had much in common. Hutchinson enjoyed Jackson's roguish sense of humour, his innate drollery. Many were the jokes made at each other's expense, and Hutchinson would tease his friend about his atrocious handwriting and his alleged familiarity with Wordsworth's poems. When Jackson died, Hutchinson proclaimed him to have been the nearest to a genius as had ever been his privilege to have known (Hutchinson 1946).

We do not ordinarily associate Jackson with the rough and tumble of medicolegal work, but at least once he appeared as an expert witness, and to the discomfiture of the cross-examining counsel. In the witness box Jackson complained he was being expected to give matter-of-fact answers to questions that were abstract. Finally, he protested "You're not asking me a question, you're merely making an epigram." The barrister blushed and sat down, muttering audibly "Epigrams? I only wish I could . . ."

Always reserved, Jackson became in later life a lonely man. He was regularly visited by his cousin Charles Jackson, who brought with him his daughter Evelyn Margaret, who died in 1985, and whose funeral I attended. Periodically he would look in at The National Hospital where the resident doctors eagerly showed him their most puzzling cases. They hung upon his words, and one of them, Kinnier Wilson, kept a notebook in which, Boswell-like, he recorded Jackson's every remark.

Jackson never attended medical lectures, theatres, or concerts. Rarely, if ever, did he dine out, and far less often did he entertain at home in Manchester Square. Those who were on terms of intimacy were few in number, but they were unequivocally loyal. One such was his colleague Thomas Buzzard. Although 4 years older than Jackson, he had been elected to the staff of The National Hospital 5 years later. His early medical career had not been in neurology. As a general practitioner in London's West End, he was involved in an epidemic of cholera in Soho. He had given evidence at the criminal trial of the alleged poisoner, Dr Smethurst. When he was 23 years of age he took part in the Crimean War, joining the medical staff of the Ottoman Army. He recorded his experiences in detail. Years later these reports were published (Buzzard 1915), for Buzzard, although modest and unassuming, was a skilful journalist.

Buzzard's visits kept Jackson in touch with what was going on in the world outside and particularly in the corridors of their beloved National Hospital. A long drawn-out dispute had been smouldering between the medical staff and the Board of Governors. In the end there was a show-down. Buzzard, known as the Nestor amongst his colleagues, was largely instrumental in securing the triumph of the doctors, and subsequent harmony. No doubt Jackson kept well out of the ring, yet he must certainly have enjoyed listening in his study to Buzzard's account of the troubles.

In conclusion, we may ask, "What emerges?" I believe we can discern a clinical philosopher who was a good man and a wise man. His virtuousness was his Northern birthright, but perhaps his Celtic forebears gave him his genius. His qualities endeared him to his colleagues. Although they did not fully understand his message, they could not but be aware of his intellectual grandeur. In his lifetime he became a legend. Even William Gowers publicly proclaimed him as master. As Anatole France wrote, "slowly but surely, humanity realises the dreams of the wise". When Jackson reached the age of retirement The National Hospital took an unheard-of step, and persuaded him to remain on the staff for another 10 years.

At the time of his death, his young colleague, Risien Russell declared, "I count myself most fortunate that I was privileged to study under such a master; the personal debt I owe him has been too much for me to hope to repay, except by lasting gratitude."

Those who seek an epitaph for Jackson will find it in Virgil: "Happy is he who can search out the causes of things, for thereby he masters all fear, and is throned above fate".

References

Acton HB (Harry Burrows) (1985) Herbert Spencer. In: Encyclopaedia Britannica, XI: 83, 3a. Encyclopaedia Britannica, Chicago

Buzzard T (1915) With the Turkish Army in the Crimea and Asia Minor. Murray, London

Duncan D (1908) The life and letters of Herbert Spencer. Methuen, London, p 227

Greenblatt SH (1965) The major influences on the early life and work of John Hughlings Jackson. Bull Hist Med 39:346–376

Hutchinson H (1946) Jonathan Hutchinson: life and letters. Heinemann Medical, London

Chapter 3

Hughlings Jackson and European Neurology

C. W. Wallesch

In John Hughlings Jackson's time neurology emerged as a subspecialty of medicine. Jackson himself, being almost entirely dedicated to the brain sciences, contributed heavily to this development, and in a number of aspects his work was of central importance.

The present study will be limited to Jackson's interactions with brain sciences and scientists on the European continent. The topic will cover three aspects: the direct and indirect influences upon Jackson, his position among European neurologists during his lifetime and how he was perceived by them, and the influences he exerted after his death.

Jackson studied medicine in York, which, at least by European standards, seems to have been a very intimate medical school. Wetherill (1961) estimated that there could have been only about a dozen students in Jackson's year, so he must have received almost private tuition. Among his teachers and tutors was Thomas Laycock, who had studied in Paris and had received his doctorate in Göttingen. It has been suggested by Bramwell (1935) that Laycock may have had the most enduring influence of all his teachers upon Jackson. Laycock was widely read in the French and German medical literature. Greenblatt (1985) considered him to be a direct link between the French tradition of anatomical diagnosis and Jackson's later work. Furthermore, Dewhurst (1982) pointed out Laycock's interest in early German and Austrian works on reflexology and aspects of consciousness, (e.g. in the works of Johann August Unzer and Georg Prochaska he translated; Laycock 1851) and implies a connection to Jackson's later fields of interest.

Among Jackson's later teachers, Charles Edouard Brown-Séquard was certainly very cosmopolitan; "a wandering scholar if ever there was one" (Greenblatt 1965). Brown-Séquard, together with Laycock, probably channelled Jackson's interests into the field of neurology (Greenblatt 1965). As Jackson was to be later, Brown-Séquard was deeply committed to the study of epilepsy. Furthermore, he was both physician and physiologist, combining clinico-anatomical and physiological approaches – a method of analysis which later became so characteristic of Jackson's work.

It is impossible to analyse Jackson's thoughts and their effects while leaving aside the influences upon him of the theories from natural and mental philosophy which he incorporated into his concepts. The central philosophical influence upon

Jackson was, of course, Herbert Spencer. Spencer, as has already been pointed out (Miller 1985), on the one hand stood in a very British tradition as an associationist philosopher, but on the other hand he espoused an evolutionism which was definitely alien – "Wagnerian" (Miller 1985).

Spencer's earliest publications dealt with phrenological problems. Young (1970) has demonstrated on the basis of Spencer's writings that "Spencer's general theory of evolution and the biological, evolutionary basis of his psychology grew out of the arguments for specialisation of functions which he elaborated in the context of his phrenological interests". There is certainly no reason to deride Franz Josef Gall and his phrenological theory today, Gall after all being the first to correlate areas of the cortex with brain functions.

In the second half of the nineteenth century the doctrine of evolution was being debated all over Europe. Spencer's own theories resembled those of Lamarck rather than those of Darwin. By 1860 the stage was set for Jackson; the concepts of evolution, dissolution and sensorimotor association were universally available. The fourth major law governing Jackson's theories, the doctrine of concomitance of physiological and psychological processes, seems, at least in its strong form, to have been stated by himself, although Jackson gives reference among others to Max Müller, Du Bois Raymond and, tentatively, to Leibniz.

In Jackson's exchange of ideas as a neurological writer with his European colleagues the French connection prevails. French was probably the only foreign language Jackson could read fluently. This might have biased his reading but does not fully explain the preponderance of the French literature in his references. There are almost 1000 quotations of facts and ideas in Taylor's (1931/32) edition of Jackson's selected writings (Table 3.1). Of these, about 20% are to the French and 10% to the German literature. The numbers of French and German authors Jackson quoted were almost equal: 44 and 46. In comparison with his contemporaries, Jackson's awareness of the international literature is exceptional.

Table 3.1. Native countries of authors quoted by Hughlings Jackson.[a] Total: 956 references

Great Britain	657
France	183
Germany/Austria	101
United States	9
Netherlands	4
Italy	2

[a] Data from Taylor (1931/32).

Among those 24 authors Jackson quoted more than 10 times, 7 are French – Broca, Charcot, Duchenne, Francois-Franck, Pitres, Trousseau and Vulpian – but only one is German – Hitzig (Table 3.2). Other Europeans Jackson gives reference to are, for example Donders, von Gräfe, Griesinger, Helmholtz, Haeckel, Leibniz, Marchi, Virchow and Wundt; this demonstrates the scope of his reading. At least 4 of the authors Jackson quoted more than 10 times were early neurophysiologists – Ferrier, Francois-Franck, Hitzig and Vulpian, which indicates the strength of his physiological interests. This interest proved to be

Table 3.2. List of authors quoted at least 10 times by Hughlings Jackson[a]

Author	No. of quotations
Anderson	10
Anstie	14
Bain	18
Beavor	15
Broadbent	24
Broca	11
Brown-Séquard	11
Charcot	10
Duchenne	11
Ferrier	49
Francois-Franck	11
Gowers	25
Hitzig	15
Horsley	16
Laycock	15
Lewes	11
Mercier	14
Moxon	12
Pitres	10
Robertson	13
Spencer	77
Todd	17
Vulpian	12
Trousseau	11

[a]Data from Taylor (1931/32).

mutual. Jackson, who never performed a physiological experiment, is today among the most frequently quoted researchers of his time in neurophysiological texts.

In the contemporary European journals Jackson was only infrequently referred to. In the first 8 volumes of *Revue Neurologique* [1893–1900] Jackson is mentioned 15 times, less frequently than, for example, Sigmund Freud's neurological writings. Reference to Jackson is most often made in the context of "l'epilepsie jacksonienne", but mainly to cases and not to interpretations. The *Deutsche Zeitschrift für Nervenheilkunde* of the same period quotes Jackson 16 times and on a broader scope of subjects (e.g. on aphasia, ophthalmoplegia and thalamic softenings), but again with reference only to cases. In both journals very few papers were published on brain diseases and their symptoms, the major fields of interest of continental neurology at that time being the spinal cord, peripheral nerve, and endocrinological and syphilitic diseases.

At the end of the nineteenth century Charcot was widely regarded as the leading European neurologist. When he died in 1893 virtually every neurological and many other medical journals published obituaries, and the German neurologists raised money for a monument. When Jackson died in 1911 none of the major continental journals even mentioned his death. With regard to the general perception of Jackson's theories in the last decades of the nineteenth and the first of the twentieth century, von Monakow and Mourgue (1928) state that they were little understood and consequently not considered at all.

Although Jackson never became much referred to in the journals, he was frequently quoted on critical arguments by a considerable number of eminent authors. Therefore, Jackson's influence upon European brain sciences has to be measured by quality, not by quantity.

Within 3 years of Jackson's death two books appeared which still influence neurology today. In 1913 Arnold Pick, Professor of Psychiatry at the University of Prague, published his *Die agrammatischen Sprachstörungen* (the agrammatical disturbances of language). Pick pointed out the implications of the influence of linguistics and psychology (e.g. the works of Wundt, Bühler, Binet, Owen and William James) upon the theory of aphasia. As Jackson, with whom he had kept up a correspondence, Pick was able to demonstrate the convergence between philosophical, psychological and neurological arguments. He could further substantiate on the basis of his observations Jackson's theories, including the doctrine of concomitance, the level structure and the laws governing evolution, and dissolution. Pick's book is the first detailed study of the linguistic aspects of the breakdown of language and thus an epoch-making event in the history of neuropsychology and behavioural neurology. Pick dedicated his book to Hughlings Jackson, his most frequently quoted author, as the "deepest thinker in neuropathology of the last century".

Constantin von Monakow, Professor of Brain Anatomy at the University of Zürich, published in 1914 his monumental *Die Lokalisation im Grosshirn* (localisation in the forebrain), in which he developed a theory of the time course of cerebral symptoms including his famous concept of diaschisis. Von Monakow's ideas concerning the interactions between different structures of the brain in conditions of disease resemble Jackson's views. However, Jackson was only quoted in the theoretical introduction. In later years von Monakow insisted that he developed his ideas for a large part independently of Jackson, having been unfamiliar with his work for a long time (von Monakow and Mourgue 1928). Von Monakow made his position towards Jackson clear in his last major work, *Introduction Biologique à l'Etude de la Neurologie et de la Psychopathologie, Integration et Desintegration de la Fonction*, which he published together with R. Mourgue in 1928. Except for Monakow himself and, notably, Freud, Jackson is the most frequently quoted author. Von Monakow and Mourgue credit Jackson with the introduction of the dynamics of growth and evolution into the neurosciences. They accept and extend Jackson's levels of consciousness, his positive and negative symptoms and the governing principles of evolution and dissolution. Von Monakow and Mourgue disagree with the doctrine of concomitance and favour an interaction between metaphysical and neurobiological processes. They introduce the vitalistic element of "horme", a carrier of instincts, which later gained importance in "neojacksonian" psychiatry, although being quite alien to Jackson's – in von Monakow's and Mourgue's view "mechanical" – concepts.

The theories of both Jackson and von Monakow include principles of the interaction between brain structures. For that reason the views of both of them have recently been discussed with increasing frequency in the neuropsychological and neurophysiological literature.

In particular, the influence of Jackson's work upon human brain physiology proved to be immense. The German *Handbuch der Physiologie* of Bumke and Foerster, which was published in the 1930s, was probably the high point of German neurological sciences. In its 18 volumes Jackson is quoted 125 times, which almost equals the number of references to Charcot. Most of the references to Jackson were made by Foerster himself, who wrote the sections on motor and sensory physiology of the brain.

Otfried Foerster [1873–1941] is a unique figure in German neurology, being neurologist, neurophysiologist and autodidact neurosurgeon in one. He was the neurologist who treated Lenin in his last years; this may indicate the esteem in which he was held. To Foerster fell the honour of delivering the lecture in commemoration of Hughlings Jackson's centenary during the meeting of the Second International Neurological Congress in London in 1935. His perception of Jackson can be best demonstrated by quoting from his talk (Foerster 1936):

> Hughlings Jackson was the first to point out that there is such a thing as the motor cortex. He was the first to state clearly that the brain, the organ of mind, possesses motor functions.

He continued:

> Nowadays the knowledge of the focal subdivisions of the precentral convolution is common to all neurologists, but Jackson had the same conception as we have to-day . . . Jackson did not use the electrical experimental method, he analysed with his keen eyes and his ingenious intuition experiments made by disease on the human brain . . . The map of the human motor cortex to be found nowadays in every textbook is nothing else but a copy of the picture engraved in Hughlings Jacksons brain.

Foerster concluded:

> The doctrines he developed have been and will be forever a guidepost for subsequent research. There is scarcely a single neurological problem which was not illuminated and elucidated by his prophetic ingenuity. Jackson's writings are the bible of neurology, the canon for the votaries of our science.

Foerster and his many disciples laid the foundation for the esteem Jackson was and still is held in by modern neurophysiologists. In Foerster's time it was demonstrated that Jackson was correct in his theories on the horizontal organisation of the cerebral cortex. The implications of Jackson's ideas on the vertical organisation of the brain are just being explored at present.

Jackson's more psychological and philosophical ideas, his doctrines of concomitance, of levels of consciousness, of the sensorimotor basis of thought, and his views of evolution and dissolution as general principles were subject to a very different line of development. Both Spencer and Jackson were already influencing French psychology in the 1880s. Their ideas exerted impact especially upon the views of Theodule Ribot, Professor of Experimental and Comparative Psychology at the College de France and predecessor and teacher of Pierre Janet. Ribot wrote a number of influential books on psychopathology, quoting Jackson extensively and approvingly. In later works, he went beyond Jackson's views in applying the concept of dissolution to intrapsychic processes, especially to neurotic regression (Delay 1953). It has been speculated that Ribot, whose books were quite popular in the 1880s and 1890s, may have influenced Freud, when the latter worked with Charcot (Dewhurst 1982).

Whereas Jackson rejected the idea of an "unconscious" level of consciousness, Ribot introduced it into his psychopathology to include primitive emotions, impulsive tendences and instincts. He thus got very close to the views held by von Monakow and Mourgue (1928). Ribot in a sense bridges the gap between Jackson's original ideas and that of neojacksonian psychiatry, e.g. of Henry Ey (for a description of Ey's theories, see Evans 1972).

Sigmund Freud started his scientific career as a neuropathologist and neurologist. As a neurologist he was strongly influenced by Jackson's work, which is especially evident in his books on childhood diplegia and aphasia (Freud 1891, 1893). In his work on diplegia Freud applied Jackson's theory of levels in the motor system and his concept of dissolution and interpreted the deficits as the consequence of deficient central inhibition. Concerning aphasia, Freud wrote (Freud 1891, p. 89, translation by Stengel 1953):

> In assessing the functions of the speech apparatus under pathological conditions we are adopting as a guiding principle Hughlings Jackson's doctrine that all these modes of reaction represent instances of functional retrogression (disinvolution) of a highly organised apparatus, and therefore correspond to previous states of its functional development. This means that under all circumstances an arrangement of associations which, having been acquired later, belongs to a higher level of functioning, will be lost, while an earlier and simpler one will be preserved.

Freud excluded his more than 40 pre-psychoanalytical scientific publications from his collected writings, and in his psychoanalytical papers quoted Jackson only twice on minor problems. Did Jackson's influence on Freud suddenly cease?

In a recent study, Schott (1985) pointed out parallels between Freud's neurological and psychoanalytical concepts, the language apparatus being a model for Freud's views of intrapsychic processes and disturbed association in aphasia, and deficient inhibition in diplegia serving as sources for elements of his theory of hysteria. Freud's writings on childhood cerebral paresis (Freud 1893), on aphasia (1891) and his first writings on hysteria (Freud and Breuer 1893, 1895) date from the same years. The assumption seems plausible that the two lines of thought have influenced each other.

Jackson's concept of brain functions, so much praised in his neurological writings, must have suited Freud's psychoanalytical theories well. The doctrine of concomitance opened the opportunity to study psychic processes without anatomical and physiological constraints. Freud was able to write (1917, author's translation): "I know of nothing less relevant to me concerning the psychological understanding of anxiety than the knowledge of the neural pathways, on which its excitations are transmitted." Freud set against the physical determinants of brain function psychological quantities such as the sum of excitement or the amount of affect, thus filling the psychological side of Jackson's model of concomitance by developing the analytical concept of energy. Additional points of contact between Jackson's and Freud's views are the concept of the positive symptom as a result of a process directed at a restoration of function, the application of principles of evolution, and the internal organisation of the psychic apparatus in levels.

The essence of this analysis of Hughlings Jackson's influences upon the brain sciences on the European continent is that whereas he may be regarded as the father of British neurology, he left a different trace in Europe; there his strongest legacies remain in the fields of neurophysiology, neuropsychology and psychiatry.

References

Bramwell E (1935) Hughlings Jackson Centenary. A commemorative dinner. Br Med J I:769–770
Delay J (1953) Etudes de psychologie medicale. Presses Universitaires de France, Paris
Dewhurst K (1982) Hughlings Jackson on psychiatry. Sandford, Oxford

Evans P (1972) Henri Ey's concepts or the organization of consciousness and its disorganization: an extension of Jacksonian theory. Brain 95:413–440
Foerster O (1936) The motor cortex in man in the light of Hughlings Jackson's doctrines. Brain 59:135–159
Freud S (1891) Zur Auffassung der Aphasie. Eine kritische Studie. Deuticke, Leipzig
Freud S (1893) Zur Kenntnis der cerebralen Diplegien des Kindesalters (im Anschlusse an die Little'sche Krankheit). Deuticke, Leipzig
Freud S (1917) Vorlesungen zur Einführung in die Psychoanalyse. Gesammelte Werke Bd XI. Fischer, Frankfurt
Freud S, Breuer J (1893) Über den psychischen Mechanismus hysterischer Phänomene. Gesammelte Werke Bd.I. Fischer, Frankfurt, pp 81–98
Freud S, Breuer J (1895) Studien über Hysterie. Gesammelte Werke Bd I. Fischer, Frankfurt, pp 75–312
Greenblatt SH (1965) The major influences on the early life and work of John Hughlings Jackson. Bull Hist Med 39:346–376
Laycock T (1851) The principles of physiology, by John Augustus Unzer, and a dissertation on the functions of the nervous system by Georg Prochaska. The Sydenham Society, London
Miller J (1985) Repression and inhibition. Paper read at the Hughlings-Jackson Symposium, The London Hospital, 30–31 October 1985
Pick A (1913) Die agrammatischen Sprachstörungen. Springer, Berlin
Schott H (1985) Zauberspiegel der Seele. Vandenhoeck & Ruprecht, Göttingen
Stengel E (1953) S. Freud's "On aphasia" (1891) London. Quoted by Dewhurst (1982)
Taylor J (ed) (1931/32) Selected writings of John Hughlings Jackson, vols 1 and 2. Hodder and Stoughton, London. Reprinted (1958) Basic Books, New York
von Monakow C (1914) Die Lokalisation im Grosshirn. Bergmann, Wiesbaden
von Monakow C, Mourgue R (1928) Introduction biologique à l'étude de la neurologie et de la psychopathologie, integration et desintegration de la fonction. Alcan, Paris
Wetherill JH (1961) The York Medical School. Med Hist 5:253–269
Young RM (1970) Mind, brain and adaptation in the 19th century. Clarendon, Oxford, p 161

Chapter 4

Hughlings Jackson and American Neurology

R. J. Joynt

The time of Hughlings Jackson in England was coeval with a time of great social change in the United States. It was also a period of great change in American medicine and during this time neurology was beginning to emerge as a specialty.

The United States had just gone through the Civil War. This had enveloped the country and had torn family and friends apart. Hundreds of thousands were killed or wounded. Usually a battle wound of any significance then meant death from gangrene or a crude amputation. The aftermath influenced American life and politics for decades.

Despite the great hiatus in normality occasioned by the Civil War, social and intellectual changes were rapidly taking place. This was also a time of worldwide change. In France there was a fairly stable government after 1848 and there was a peaceful interlude until the Franco-Prussian War. Great science was being produced by Pasteur and others. Britain was basking in the glory of the Victorian Age. While the British had been fighting "little wars" – if the Crimean campaign and the Sepoy rebellion can be considered little – they were prospering with great new ideas such as those of Wallace and Darwin.

In the United States, there was extensive westward expansion and the east and west were connected by rail in 1869. Many of the great universities, public and private, were being established, and the concentration of intellectual power in the few eastern private schools was being dispersed. There was an eagerness to learn and popular philosophers abounded. Herbert Spencer, Hughlings Jackson's favourite, was also a favourite in the United States. Any home espousing to intellectual pursuits had a copy of *First Principles* prominently displayed, perhaps with the pages uncut.

The American Medical Association had been founded, so there was some semblance of organisation within the medical profession. Medical journalism had begun. Medical education, however, was extremely uneven. Although some of the good universities had medical schools, much of the education was still done by preceptorship in a doctor's office or in poorly equipped and poorly staffed institutions which called themselves medical schools. There was a great proliferation of these until curtailed by the reactions of the public and the regulations by the states following Abraham Flexner's report in 1911.

While neurology was emerging as a specialty in America, it did not take the same form as in western Europe and Britain. The great era of clinical pathological correlation taking place in Germany and France did not occur in America. To some extent, in Germany at least, part of this impetus arose from the great psychiatric institutes, where structural changes were being sought to explain mental aberration. These institutes spawned a number of eminent neurologists with knowledge of pathology and anatomy. The chief contributions in Britain, and most notably that of Hughlings Jackson, were the physiological insights into the normal and diseased nervous system. In Spain and Italy, scientists working on the nervous system were dealing with the cellular organisation, as revealed by their ingenious histological methods.

The United States did not have these foundations, and the early pioneers in neurology were clinical neurologists. And while we would classify Hughlings Jackson as a clinical neurologist, he always thought in terms of the physiology of normal and disrupted function. Thus, the most notable leaders in American neurology were Silas Weir Mitchell, William Hammond and James Jackson Putnam. They made many contributions, but their greatest were in the clinical descriptions of various disorders. There were some minor exceptions. Hammond, for example, did deal with the problem of aphasia. James Jackson Putnam toward the end of his career incorporated psychoanalytic principles into his neurological thinking.

As a consequence of the type of practitioner and the type of intellectual approach to the nervous system, Hughlings Jackson was, in fact, rarely alluded to in most of the early writings of the American neurologist. There is no doubt, however, that he was very highly regarded in the United States.

Silas Weir Mitchell dedicated his book, *Lectures on the Diseases of the Nervous System, Especially in Women*, published in 1881, and his *Clinical Lessons on Nervous Diseases*, published in 1897, to Hughlings Jackson. The dedication in the former book was "Dedicated to J. Hughlings-Jackson, M.D., F.R.S. With warm personal regard and in grateful acknowledgement of his services to the science of medicine." Interestingly, Hughlings Jackson is not referred to in the text in that book. Mitchell referred to Hughlings Jackson as "my old friend" and commented on one occasion ". . . it has always surprised me that he was not rewarded or that a title was not given him by the government, while in my time a dozen or more already forgotten have been made baronets". Mitchell's great contributions, of course, were in an area of neurology different from Hughlings Jackson's. His most important work came in the study of injuries to the peripheral nerves during the Civil War. His experience at Turners Lane Hospital in Philadelphia was remarkable, and his descriptions of causalgia and similar conditions are still classic contributions.

Weir Mitchell was also familiar with Hughlings Jackson's interest in the ophthalmoscope, and in a paper in 1873 with William Thompson he states, "the ophthalmoscope becomes of imperative use in treating nervous diseases, and should be employed in the examination of all cases as part of a routine, as advised by Mr. Hughlings Jackson whether the patient complains of any defective sight or not".

In the first American textbook of neurology, *A Treatise on Diseases of the Nervous System*, published in 1871 by William A. Hammond, Hughlings Jackson's contributions to neurological observations are mentioned several times. Hammond was a controversial figure in American medicine. Early in his career he joined the Army, then resigned, and eventually rejoined during the Civil War and rose to be Surgeon-General. He was only in his mid-thirties when he achieved this position and his meteroric rise irritated others in the Army and Government. He was eventually court-martialled but was reinstated several years later. After he left the Army in 1863, he became a practising neurologist and psychiatrist in New York City and was professor at the University of the City of New York and at Bellevue Hospital Medical College. He had met and in fact collaborated with Weir Mitchell on earlier occasions. His textbook in 1871 was the first textbook of neurology in the English language. It got variable reviews, including one in the *Medical Record* which stated, "When he is right he is clearly right, when he is wrong he is clearly wrong." It was pointed out by some that much of what Hammond wrote was in the lectures by Charcot. Hammond refers to Hughlings Jackson in several places, but not in the section on epilepsy. Hammond had a particular interest in aphasia and his chapter in the book has a very extensive history of thought on that disorder. He cites several of his own cases to back up his contention that the organ of language is situated in both hemispheres, particularly that part of the hemisphere that is nourished by the middle cerebral artery. However, he does go on to say there is evidence to show that the left side of the brain is more intimately connected with the faculty of speech than the right. He notes that the more frequent occurrence of aphasia with right hemiplegia is due to the fact that a cerebral embolus is more likely to plug the left middle cerebral artery. In this contention, he states the following, "Dr. Hughlings Jackson has very satisfactorily worked out the relation and my own experience, presently to be related, abundantly confirms the fact". In none of the references to Hughlings Jackson in the book does he utilise Hughlings Jackson's thoughts on the altered physiology, but refers only to Jackson's observations on pathology.

A slightly less well-known but nonetheless important neurologist was James Jackson Putnam. Putnam was one of the seven original founders of the American Neurological Association in 1875 and was its President later. It is likely that Putnam's work was influenced more by Hughlings Jackson than any of the early neurologists in this country. Putnam graduated from Harvard Medical School and in 1870 did the traditional trip to study on the Continent and in Britain. In England he met and studied under Hughlings Jackson. He always considered Hughlings Jackson as one of his most important mentors. Following this he set up practice and was particularly interested in electrotherapy. He started a neurological clinic at Harvard Medical School in 1874 and in 1893 he became Professor of Diseases of the Nervous System at Harvard, holding this position until his retirement in 1918. Putnam was very much influenced by Hughlings Jackson's teaching that disease of the nervous system was the result of the imbalance of excitatory and inhibitory states. He used this theme throughout his neurological career. He made many original observations on various diseases including poliomyelitis, polyneuritis, degeneration of the spinal cord and heavy metal poisoning. Later on he had an interest in the diseases of the basal ganglia.

Late in his career, Putnam became influenced by Freud and had an active correspondence with him. When Freud came to the United States for his famous visit to Clark University in 1910, he was very much taken with Putnam. Freud stated at that time,

> The most important personal relationship which arose from the meeting at Worcester was that with James J. Putnam, Professor of Neuropathology at Harvard University. Some years before he had expressed an unfavourable opinion of psychoanalysis, but now he rapidly became reconciled to it and recommended it to his countrymen and colleagues in a series of lectures which were as rich in

content as they were brilliant in form. The esteem he enjoyed throughout America on account of his high moral character and unflinching love of truth was of great service to psychoanalysis and protected it against denunciations which in all probability would otherwise quickly have overwhelmed it. Later on, yielding too much to the strong philosophical bent of his nature, Putnam made what seemed to me an impossible demand – he expected psychoanalysis to place itself at the service of a particular moral-philosophical conception of the Universe – but he remains the chief pillar of the psychoanalytic movement in his native land.

Indeed, Putnam tried to weave his Jacksonian conception of the nervous system into what he learned from his studies on psychoanalysis. An excellent example of this was in his essay, "On the Etiology and Pathogenesis of the Post-traumatic Psychosis and Neuroses", which he published in the *Journal of Nervous and Mental Diseases* in 1898. Here he notes that Hughlings Jackson looked upon the healthy nervous system as an equilibrium between various degrees of excitation and inhibition, and that in disease this equilibrium was disturbed. However, Putnam added to these psychic forces that could play a role in this balance. He gratuitously excuses Hughlings Jackson with the following statement, "what Dr. Jackson, could not then realize for lack of sufficient data, was that, in this readjustment, subconscious processes of a high order and susceptibility of study and classification greatly complicated the situation, making our analysis at once more difficult and more statisfactory". It appears that, among the early neurologists, it was James Jackson Putnam, who personally trained under Hughlings Jackson, who was most struck with Jackson's views of the nervous system.

Another great admirer of Hughlings Jackson was William Osler. Osler, trained at McGill, had been Professor of Medicine at the University of Pennsylvania and Johns Hopkins. A lesser known fact is that Osler was an honorary member of the American Neurological Association. In 1901 Osler, along with Silas Weir Mitchell and James Jackson Putnam, requested that Hughlings Jackson publish a collection of his papers. Hughlings Jackson declined at that time, and it was not until three decades later that publication was to be accomplished. In a letter in 1904, Osler alludes to a literature search he had made for some diagnostic point by Hughlings Jackson, so it was obvious he was quite familiar with Hughlings Jackson's writings. On 4 September 1911, in a letter to Frederick C. Shattuck, Osler singles out Hughlings Jackson as someone who has not been honoured properly. This was just after Osler was knighted. Osler wrote, "some of the most deserving men never receive any recognition – Hughlings-Jackson for example and imagine chucking at 80 a knighthood to Jonathan Hutchinson!"

Hughlings Jackson was also recognised in the non-medical world. William James, Professor of Psychology at Harvard, published in 1891 his *Principles of Psychology*. He agrees with Hughlings Jackson's views on the parallelism of brain and mind. James notes that the brain and the mind alike consist of simple elements, sensory and motor. In this he quotes Hughlings Jackson, "all nervous centers from the lowest to the very highest (the substrata of consciousness), are made up of nothing else than nervous arrangements, representing impressions and movements . . . I do not see of what other materials the brain can be made". James' book was probably one of the most influential forces in American psychological thought.

In summary, then, it is evident that Hughlings Jackson was widely read and highly regarded in the United States. His clinical and pathological observations were frequently cited. However, at least early on, his views on the organisation of the nervous system were less well used. This probably reflects the interests of the few people in the specialty of neurology at that time. These did gain popularity and were important in Putnam's work on psychoanalysis and in William James' psychological thinking. Hughlings Jackson was made an honorary member of the American Neurological Association in 1881. It was a most august group that year – Hughlings Jackson, Charcot, Erb, Westphal, and Meynert. It was also true that the egalitarianism loudly proclaimed in America did not override the mystification of the Yankees as to why he was not knighted. This peculiar American schizophrenia, however, was explained by one of my English colleagues, who said that no one so loves an aristocrat like a democrat.

References

Hammond WA (1871) A treatise on diseases of the nervous system. Appleton, New York

James W (1891) Principles of psychology, vols 1 & 2. Macmillan, London

Mitchell SW (1881) Lectures on the diseases of the nervous system, especially for women. Lippincott, Philadelphia

Mitchell SW (1897) Clinical lessons on Nervous diseases. Lea, Philadelphia

Putnam JJ (1898) On the etiology and pathogenesis of the post-traumatic psychosis and neurosis. J Nerv Ment Dis 25:769–799

Chapter 5

The Hughlings Jackson Tradition at The London Hospital

R. A. Henson

Hughlings Jackson was already a lecturer in pathology at The London Hospital when he was elected fourth assistant physician on 25 August 1863. Both appointments had been engineered by Jonathan Hutchinson [1828–1913], a celebrated physician and surgeon with some 1500 communications to his name, including several primary neurological observations. Jackson's resignation was accepted on 25 February 1894 after the customary 30 years on the senior staff. His portrait was painted at this time and still hangs on the walls of the Medical College, a measure of the regard in which he was held.

The neurological history of The London Hospital had already begun with James Parkinson [1755–1824], an early student at The London. Apart from his account of the Shaking Palsy in 1817, Parkinson was a man with wide interests: radical in politics, he actively demonstrated his sympathy with the French Revolution, and he was also an amateur geologist of some distinction.

Parkinson was followed by a group of eminent Victorians, broadly contemporary with Jackson: W. J. Little [1810–1894], who described spastic diplegia and founded the Royal Orthopaedic Hospital; J. L. H. Langdon-Down [1828–1896], of Down's syndrome; Jonathan Hutchinson and his son, Jonathan Jr., who described congenital fusion of the cervical vertebrae in 1894; Warren Tay [1844–1921], of Tay–Sach's disease; and Sir William Lister [1866–1941], the ophthalmologist. Brain (1959) reviewed the contributions of these men and their successors in the twentieth century.

Brain discerned three main strands of continuity from Jackson onwards at The London Hospital: first, an approach to neurology through physiology; second, a dynamic and holistic attitude; and, third, close links between neurology and general medicine. Henry Head [1861–1941], George Riddoch [1889–1947], and Brain himself [1895–1966] were the main repositories of the Jacksonian tradition within the Hospital. Riddoch made important studies on injuries of the spinal cord and occipital lobe of the brain during the First Word War, but he was eventually overwhelmed by the demands of practice, to the detriment of research. Nevertheless he remained true to the physiological philosophy which he had inherited from Jackson through Henry Head, whose pupil he was. Brain was a polymath whose contributions to medicine and medical politics, literature and philosophy, are well remembered 20 years after his death. Henry Head, however, was the most interesting figure of the three in terms of interests, ideas and originality of thought. Possessed of an attractive temperament and talented in many fields, his influence was wide, both within and without medicine. He always acknowledged his indebtedness to Jackson, and the following account of a visit to Jackson's house appears in his private papers.

> The door of a small house in the corner of Manchester Square is opened by a very old butler who says he will see if the doctor can see me – "he has not been well lately but is better now". I am ushered into a large room in which my eye is at once attracted by two immense arm chairs. By each of them stands a low table with notebooks and pencils. Roundabout, in apparently hopeless disorder, lie innumerable books, pencilled journals and scraps of typewritten manuscript. From one of these chairs rises a white haired man in a wrinkled old fashioned frock-coat. Standing with his head a little on one side like a giant bird, he stretches out to me a most friendly hand. His words come thickly through the veil of his heavy white beard and moustache which entirely cover his mouth. I draw close to his great chair and plunge into the middle of a scientific conversation. At first he is somewhat shy for his life during many years has been lonely and this loneliness has been increased by the onset of deafness. Although nearly 70 years old he is still full of ideas and has maintained a wonderful freshness of interest. But he has a curious and embarrassing habit of assuming, in his great modesty, that fundamental principles enunciated by him a quarter of a century ago are still unknown to me, to whom they have been elemental steps in intellectual training. During our conversation he repeatedly rises to fetch some paper or book or to refer to one of his many notebooks.
>
> Under his shyness he hungers for affection and has been the fairy godfather to a multitude of children, many of them are grown up, some are married, and a host of photographs fill the mantelpiece.
>
> For recreation he reads novels – not those with pretentions to literary value but the ordinary novel of the circulating library. Stacks of such books occupy a corner of his room: but he never speaks of them. They form a pure pastime.

This record has been constructed from the mass of personal letters, notebooks and other documents which have survived since Head's death, supplemented by memories and observations of family and friends. He began an autobiography in 1926 but only completed some 50 quarto pages, covering his earlier years; evidently illness precluded completion.

Head was born on 4 August 1861, third son of Quaker parents. He was known as Harry by family and intimates, as Henry by students and others. His father was a broker at Lloyds; the syndicate prospered, according to family lore, because they were the only group who would insure residents in San Francisco against earthquake. Harry recorded his progress through his public school, Charterhouse, and Trinity College, Cambridge, in regular letters to his mother. Between leaving school and going up to Cambridge Head spent two months at Halle studying physiology and histology. Leaving home entirely ignorant of German he returned fluent, with an abiding interest in German literature. After Cambridge Harry spent 2 years with Hering at Prague, and his best laboratory work stemmed from this period. His experiences in Halle and Prague were documented in letters to his mother and in the autobiographical fragment. The picture which emerges is of a confident, independent and physically active young man.

It was in Prague that he sat in a beer-garden oblivious to the passage of time while he read the latest Ibsen play. A less serious episode concerns a visit of the younger Strauss to Prague with his orchestra. Harry, with a group of friends devoted to classical music, decided to go to a concert and boo the musician off the platform. Wearing evening dress they took the first two rows of the stalls, waiting to make their demonstration. In the event they were so captivated by the Strauss music that they ended up dancing on the stand.

Head was elected Assistant Physician at The London Hospital in 1896, by which time he had completed his seminal work on visceral sensation. In the same year he had failed to gain election at the National Hospital, in retrospect a strange happening. He had already determined to pursue the physiology of sensation, a task which engaged him for many years. Initially he studied zoster, gathering material from several hospitals, and peripheral nerve injuries, both at The London and at Netley Military Hospital. His mind was far too active and original to be content with observation and description; thus he produced his much criticised general theory of sensation. Writing about Head's research, Adrian, a pupil, said that Head "had not been given the credit he deserved"; he classed him with Sherrington and Keith Lucas as "the three men who had most influence on neurological progress in England during the classical period".

Head was dependent on private practice for his living, but this work was distasteful for two reasons: first, because he disliked charging fees; indeed, he was contemptuous of colleagues who displayed their pursuit of money; second, because he begrudged the time lost from research, and research was the true passion of his life. Although he lived frugally these attitudes delayed his marriage to Ruth Mayhew, an Oxford lady who appears in Kilvert's diaries and, with her sisters, in the letters of C. L. Dodgson (Lewis Carroll). Fortunately, Harry's father intervened with a timely annual subvention, and the couple were married in 1904, after an unusual courtship lasting 7 years; Harry was almost 43 and Ruth 37 at the time. Complete financial freedom came a few years later with the deaths of his parents. Throughout, Head had financed his research and paid his assistants. He did not believe in purchasing property and always rented his homes, both modest and magnificent. Cushing was highly impressed by the Heads' final residence, Hartley Court, when he called there in 1936, though he departed with faulty information about their courtship (Fulton 1946); it seems likely that Harry or Ruth had been teasing him.

In appearance Head was short and portly, bearded and moustached, and elegant in dress. Despite his build he was an active man, who played tennis and hockey; when past 40 he rode a bicycle race against some young Germans whom he left far behind on the road into Regensburg. He enjoyed riding – he once took over a mettlesome mare which Rivers could not control on a Cambridge excursion – and shooting, particularly with Horsley.

As a research worker Head was tireless, enthusiastic and inventive. Young men competed to serve with him in clinical and laboratory work. His position at The London was comparable to that of his friend Holmes at Queen Square between the wars.

There is ample testimony of his skill and popularity as a teacher, and stories about him abounded in times past. One tale concerns a patient chasing Head with a knife. The following account is probably nearest the truth. Late one night Sherren and Head were engaged in an unofficial post-mortem examination, assisted by a porter bearing a lantern; suddenly this man fell to the ground in a major fit. After giving aid the couple returned to their task: removal of the spinal cord. Sherren, an unwilling collaborator in the enterprise, left Head to prepare the specimen. Hearing a noise Head turned and saw the porter approaching with a long post-mortem room knife. A chase round the tables ensued; fortunately the disturbance alerted a night-porter on his rounds, and he summoned help. The epileptic man was disarmed, Head made suitable disbursements, and everyone was satisfied and subsequently silent.

Like Charcot, Henry was always pleased to teach on hysteria. One day he stopped with his train at the foot of the bed of a beautiful girl with a mass of black hair, suffering from this complaint. "These hysterical girls are often subject to erotic outbursts." he concluded "Look straight at my finger, my dear," he commanded. The next moment the patient had seized him around the neck, his upper half was enveloped by her hair, save for his beard which insisted on protruding itself; the sound of a barrage of kisses followed – with Henry's delighted voice exclaiming "Didn't I say so gentlemen? Didn't I say so? Typical, gentlemen! Oh, typical!"

Head was well-known in artistic and intellectual circles from the 1890s onwards. When he investigated spiritual healing at Lourdes at this time, one of his ecclesiastical sponsors was Cardinal Manning. Who else would have sufficed? On his German travels he was twice a guest at a ducal castle which was a resort of the German royal family. His descriptive letters show him to have been at his uninhibited best during these visits, from which he departed on his bicycle.

Harry's skills in literary, dramatic, and artistic criticism were founded on careful evaluation and detailed analysis of material; thus we find him detecting Thackeray in minor historical error and apparent plagiarism of Pope; deprecating R. L. Stevenson as a letter writer; analysing the character of Emma Bovary; discussing the art of Duse and Rejane; tutoring a future expert in Chinese art; and commenting on Heine, Maeterlinck, the decadent French poets, Henry James, Wagner, Watteau, and artists of the Italian schools who were among his passions. He was highly critical of Shaw in 1900; he felt that Shaw was wasting his genius at a time when he could have been an effective critic of British institutions and government. Yet poets came first, and he was himself a poet. Henley, Hardy, Sassoon, Graves and Nichols were all friends. Graves called him a "terrific man" and Sassoon was a life-long admirer. Friends in other fields included Berenson, Whitehead and Bertie Russell, to cite a celebrated trio. All who knew him spoke of his wisdom and encyclopaedic knowledge, but Holmes alone among biographers perceived the complexity of the man.

Harry and Ruth moved easily in society in their brilliant days, yet they were unconventional. Before marriage they walked together in London and visited galleries and theatres unchaperoned. Ruth even left her post of headmistress in Brighton early one day to meet Harry in London, and he supplied her with cigars.

Harry cared nothing for politics, though he worked for scientific and educational advance, and for The London Hospital. He addressed this quatrain on the South African War to the shade of Dr Samuel Johnson:

> Patriot is scoundrel by another name
> You cried to cap your ridicule with shame.
> Could you with your dogmatic thunder quail
> The heart of Kipling and the *Daily Mail*.

On obsequious attitudes to royalty: a great astronomer speaking at a Royal Society dinner said he knew "that at the name of the Queen our hearts are moved not only to reverence but to love". As Harry, seated next to Gowers, heard everyone around him beating the tables in acclamation he commented, "I wonder whether they or I are mad."

Head's originality of thought was manifest in many ways and in many fields, as his personal papers reveal. One example must suffice. In 1900 he wrote, "medical education in England suffers from the fact that the great hospitals are manned by practitioners of medicine who sometimes teach instead of the professors of that science who occasionally practice", a notion echoed by later educationalists both in Britain and abroad.

Although gregarious by nature, ordinary social life had no appeal for him; in such circumstances he tended to become impatient or bored. He needed the stimulus of interested or interesting people from all classes, though he naturally shone in the company of intellectuals. Adept at managing or directing conversation, he was able to delight hospital residents with readings of verse after dinner. Popular, erudite and cultured, he was nevertheless highly successful with children, for there was much of the child in him.

While Harry enjoyed female company there is no evidence of any romantic attachment apart from that with Ruth Mayhew, and he was 36 when their friendship began. He once told her of his plan to marry a young wife when his career permitted it, and he certainly showed enough interest in Vanessa Stephen to arouse Ruth's jealousy. Head and his mother were unusually close, and this fact, plus his preoccupation with work, probably accounted for his delayed pursuit of Ruth, who in turn showed signs of desperation as their long courtship continued without apparent hope of marriage. Despite indications of sensuality, as in his responses to Rivers' sensory testing and his drawings in the papers on visceral sensation, he waited 6 years before he kissed Ruth; thereafter he embraced the new relationship with enthusiasm. More broadly, he conversed with women as equals, and, at the other end of the social scale, there is written evidence of his sympathy and understanding with the poor women who thronged the outpatient department at the Hospital.

What a paragon of knowledge, scientific dedication, artistic skills and social grace! Henry Head was larger than life. Biography demands a search for flaws, and these there were. His enthusiasms could be intemperate, and this showed both in his work and literary criticism. Some colleagues found him over-combative in debate and dismissive of differing opinions. He must have been a fearsome opponent, factually informed and confident about the proper course to adopt. Until his marriage he could be impatient, unkind, and even rough in his behaviour within the family; his response to his sister's engagement to an army officer was almost violent. Ruth Mayhew accused him of being boastful and domineering in one bitter letter, but once he fell in love with her, and he took his time over that, he was changed, if not tamed.

Although apparently a severe materialist, Head was interested in certain forms of mysticism, as Holmes observed. His loss of religious faith at the age of 18 was compounded by the realisation that he had thought more deeply about the subject of prayer than Bloomfield Jackson, the clergyman engaged by his parents to reason with him. Jackson terminated the interview after 10 minutes by asking Harry to bring him a glass of sherry. Regarded by some as a classical humanist, enquiring agnostic is a more apt description. In his last years he confessed to belief "in a mystery that only time or circumstance will completely reveal".

Hughlings Jackson received no public honours. Head's knighthood came late, at a time when he was crippled and unable to work or write. He was a British nomination for the first Nobel prize in medicine [1901–1902], and knowledge of this brought him great pleasure. He observed the first sign of Parkinsonism at 56, and the next 23 years saw a gradual physical decline, faced with great personal courage. At his passing on 8 October 1940, there died one whom we may claim as the most whole man among English physicians of his time, and perhaps of the century.

This stanza from his "Spring Death" tells much about him:

But death has stripped me of all desire,
An outcast from earth's generous festival,
I go to warm me by the altar fire,
Whereat we worshipped. Happy little shrine –
Soft garlands on the wall,
The music and the laughter and the wine,
Talk, like a fountain pulsing to the blue,
To fall in rainbow droplets on the grass,
Warm human joys – they shall my heart renew,
They cannot pass.

References

Brain R (1959) The neurological tradition of The London Hospital or the importance of being thirty. Lancet II:575–581
Fulton JF (1946) Harvey Cushing: a biography. Blackwell Scientific, Oxford

Section II

CONSCIOUSNESS AND MEMORY

Chapter 6

Hughlings Jackson's Views on Consciousness

J. C. Eccles

Jackson was much more subtle in his writings than is generally believed. He was a philosopher always. In his first years he thought of devoting himself to philosophy and he kept that interest throughout life, and of course in his association with Herbert Spencer. However, I believe he finished far ahead of Herbert Spencer. He did not just follow him.

I have used as my source of Jackson's beliefs on brain and consciousness both the Croonian lectures he delivered at the Royal College of Physicians in March 1884, and his subsequent lectures on the evolution and dissolution of the nervous system. These are in Volume 2 of the collected works of Hughlings Jackson (Taylor 1931/32); this volume contains his mature thought. Earlier, I think, he was a materialist, but most people who start that way become wiser as they get older! He insisted that mental states are associated with activities of the highest levels of the nervous system, particularly the frontal lobes. This is of course very much in line with the work by David Ingvar using the radio-xenon testing to show the high activity of the frontal lobe in mental states and performance.

Jackson equates mind with consciousness, a view with which I agree. He emphatically rejects panpsychism and he finds unconscious states of mind unintelligible. When he makes statements, he says them quite sharply. I give a dualist statement from his writings in the Croonian Lecture: "If anyone wished to be thoroughly materialistic as to what is material in the nervous system, let him not be materialistic at all as to the mind which is not material at all. A man has both a mind and a body" (Taylor 1931/32, p. 63). Central to his philosophy is his concept of the doctrine of concomitance – I have a mind–brain.

In Fig. 6.1 World 1 is the world of matter and energy, including human brains. The materialist theories of the mind have to insist that World 1 is closed. The closedness of World 1, as Popper and I say, is central to their doctrine (Popper and Eccles 1985). An outside mental state (World 2) cannot act upon the World 1_m in Fig. 6.2. The various materialist theories of the mind are briefly presented in the upper four entries of Fig. 6.2. Radical materialism says that there is no mind at all! We go on to what Jackson believed. He was against panpsychism and he rejected epiphenomenalism. He had a parallelist theory of concomitance, which resembles the present identity theories in that World 1, including the brain, had special parts of the brain, World 1_m, that were "identical" with World 2, the world of the mind. That is, he retained the idea that there were mental states, the

WORLD 1	WORLD 2	WORLD 3
PHYSICAL OBJECTS AND STATES	STATES OF CONSCIOUSNESS	KNOWLEDGE IN OBJECTIVE SENSE
1. INORGANIC Matter and energy of cosmos 2. BIOLOGY Structure and actions of all living beings - human brains 3. ARTEFACTS Material substrates of human creativity of tools of machines of books of works of art of music	Subjective knowledge Experience of perception thinking emotions dispositional intentions memories dreams creative imagination	Cultural heritage coded on material substrates philosophical theological scientific historical literary artistic technological Theoretical systems scientific problems critical arguments

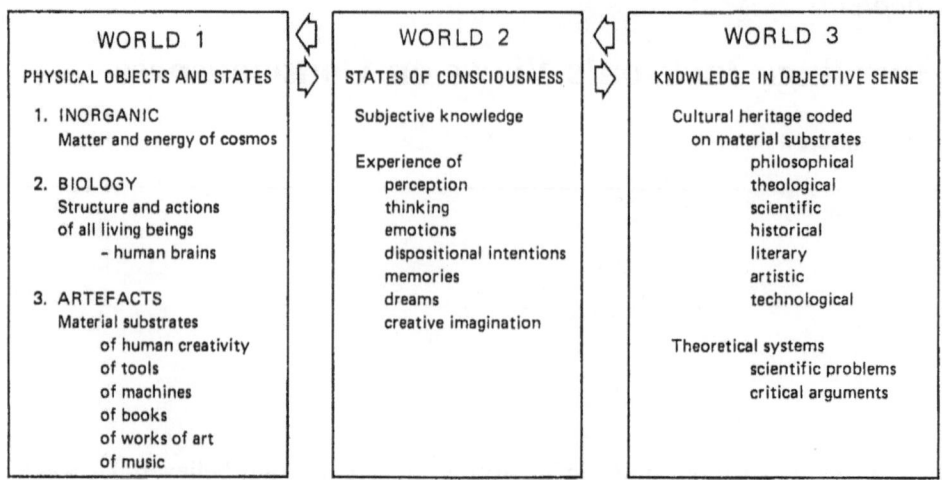

Fig. 6.1. Tabular representation of the contents of the three worlds that comprise everything in existence and in experience.

mind, but he resolved the problem of World 1 being closed, by proposing concomitance.

Jackson presented his doctrine of concomitance in three statements:

1. "States of consciousness (or synonymously, states of mind) are utterly different from nervous states" (Taylor 1931–32, pp. 72–74). This is a dualist statement.
2. "That the two things occur together – that for every mental state there is a correlative nervous state" but no one knows the nature of the relationship. Jackson got closer than most people in trying to talk about that, but he did admit the enigma.
3. "Though mental states and nervous states occur in parallel, there is no interference of one with the other". They are running parallel or are identical.

Now in the third aspect of his concomitance there was no interaction. He writes there that non-material conscious or mental events cannot cause actions because that is contrary to the doctrine of the conservation laws. He was quite advanced for his time. He knew exactly what we would say today to the closedness of World 1. He states, "those who accept the doctrine of concomitance do not believe that sensations, volitions, ideas and emotions, produce movements or any other physical states" (Taylor 1931/32, p. 86.3). It was a very clear idea that all perceived sensations do not produce movements. Then he goes on, and this is a very interesting statement, "movements always arise from liberations of energy in the outer world, and it would be marvellous if there were an exception in our brains, marvellous if, for example, the Will, an immaterial agency, interfered in the activities of nervous arrangements of the higher centres" (Taylor 1931/32, p. 86.5). That would indeed be marvellous, and my task is to tell you that his

World 1 = All of material or physical world including brains
World 2 = All subjective or mental experiences
World 1_P is all the material world that is without mental states
World 1_M is that minute fraction of the material world with associated mental states

Radical Materialism: World 1 = World 1_P; World 1_M = 0; World 2 = 0.

Panpsychism: All is World 1–2, World 1 or 2 do not exist alone.

Epiphenomenalism: World 1 = World 1_P + World 1_M
World $1_M \longrightarrow$ World 2

Identity theory: World 1 = World 1_P + World 1_M
World 1_M = World 2 (the identitiy)

Dualist – Interactionism: World 1 = World 1_P + World 1_M
World $1_M \rightleftarrows$ World 2; this interaction occurs in the liaison brain, LB = World 1_M.
Thus World 1 = World 1_P + World 1_{LB}, and
World $1_{LB} \rightleftarrows$ World 2

Fig. 6.2. Diagrammatic representation of brain–mind theories that incorporates the World 1 and World 2 of Fig. 6.1. The essential features of the materialist theories of the mind are summarised for panpsychism, epiphenomenalism and the identity theory. The last theory has a variety of names according to the whims of the creators of the minor varieties of what are essentially parallelist theories. The subdivision of World 1 into World 1_P and World 1_M helps in clarification of their specific features. World 1_M is assumed to be restricted to special areas of the brain in epiphenomenalism, the identity theory, and dualist-interactionism. The essential and unique feature of dualist-interactionism is shown by the reciprocal arrows between World 1_M and World 2 in the second line.

marvel can be found. It would, I think, have pleased Jackson considerably because he was very much insistent on this, as he was of course on the physics of the nineteenth century, as most of us still are.

Jackson's statements about movement are very interesting, although it is not again generally recognised that they are, in some respects, contrary to his concepts of concomitance. He was inconsistent, but with somebody as original as Jackson you do not expect all to be laid out in a kind of Euclidean manner. He writes about movement as follows (Taylor 1931/32, pp. 199.9–200.2): "the operation is nascently done before it is actually done. [That is, you think of the movement before you actually do the movement.] There is a 'dream' of the hand as being already put out – before I can think of now putting it out." Jackson made this statement, and I want to put it on record again that, whenever you are planning a movement, you already have it in your mind before you do it. The will he is talking of is an immaterial agent in the mind and it brings about the action. He has made no attempt to recognise a dilemma or to make an explanation of this in his parallelist philosophy.

I now introduce a modern experiment on the will. A radio-xenon injection was made into the internal carotid artery by Roland and associates, and the radiation from all the different cerebral vessels gave a record of nervous activity, which was picked up by 254 Geiger counters over that hemisphere. Fig. 6.3a shows that when the subject was carrying out a complex learned movement of his hand – the motor sequence test – but not otherwise moving, there was an increased activity of the contralateral sensorimotor hand area, as would be expected. However, there was also an increase in the supplementary motor area on both sides. The extraordinary finding, shown in Fig. 6.3b, where the subject was mentally doing the movement but not moving, was that the increased activity was only in the supplementary motor areas. Thus the story of Hughlings Jackson's "dream" is now shown to be associated with actual activity of the supplementary motor area before any motor action is carried out.

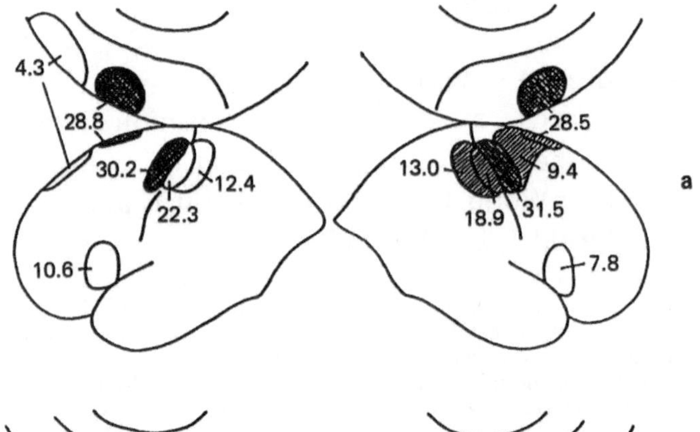

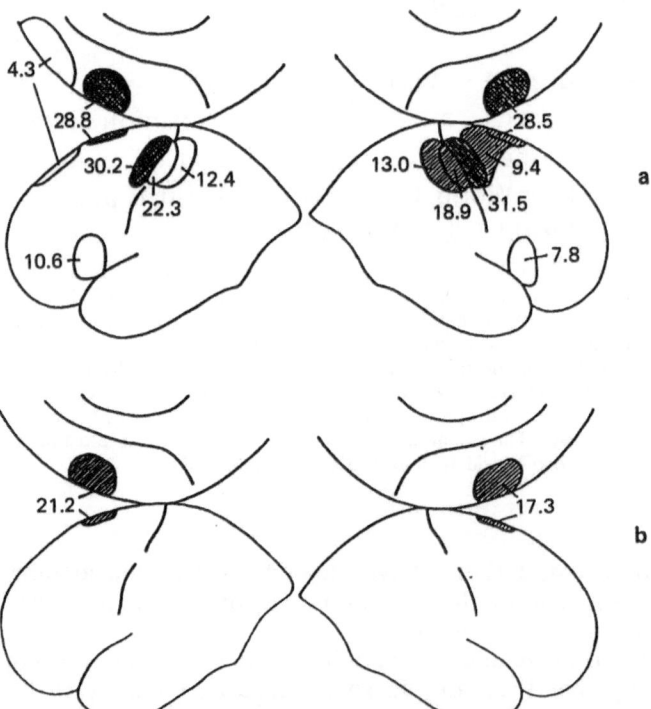

Fig. 6.3. **a** Mean percentage increase of the rCBF during the motor-sequence test performed with the contralateral hand, corrected for diffuse increase of the blood flow. *Cross-hatched areas* have an increase of rCBF significant at the 0.0005 level. *Hatched areas* have an increase of rCBF significant at the 0.005 level; for other areas shown the rCBF increase is significant at the level 0.05. *Left*, left hemisphere, five subjects. *Right*, right hemisphere, 10 subjects. **b** Mean percentage increase of rCBF during internal programming of the motor-sequence test, values corrected for diffuse increase of the blood flow. *Left*, left hemisphere, three subjects; *right*, right hemisphere, five subjects. (Roland et al. 1980).

The intention of the movement is in World 2 before you carry it out. It is shown in Fig. 6.4 acting on the supplementary motor area (SMA in Fig. 6.4) in the way that has been observed in Fig. 6.3b. Here in Fig. 6.4 we have intention to move, a mental state acting upon neurons of the supplementary motor area, and giving rise eventually to the movement. How can mental states here act upon the brain states? We now turn to Henry Margenau, a distinguished quantum physicist, who wrote a book called *The Miracle of Existence* (Margenau 1984). There he outlined a theory whereby quantum physics could possibly explain this phenomenon of brain–mind interaction which has been puzzling us for so long.

If we are going to understand how mental events can act upon neural events, we have to go down to the actual synaptic micro-sites where mental events could act upon neural events. Figure 6.5 could be an ordinary synaptic bouton on a dendrite of a pyramidal cell in the motor cortex, which has about 10000 such boutons on it. Figure 6.6 is an idealised diagram of a bouton to show the special arrangement of synaptic vesicles in the presynaptic vesicular grid. For any one bouton, there are about 50 vesicles on the firing line, all arranged in this paracrystalline manner.

The second important item of information comes from the work of Redman and associates, who have shown that, whenever an impulse comes to a bouton, only now and then does one vesicle come out, never more than one, the mean probability varying widely for different boutons. For the whole ensemble the

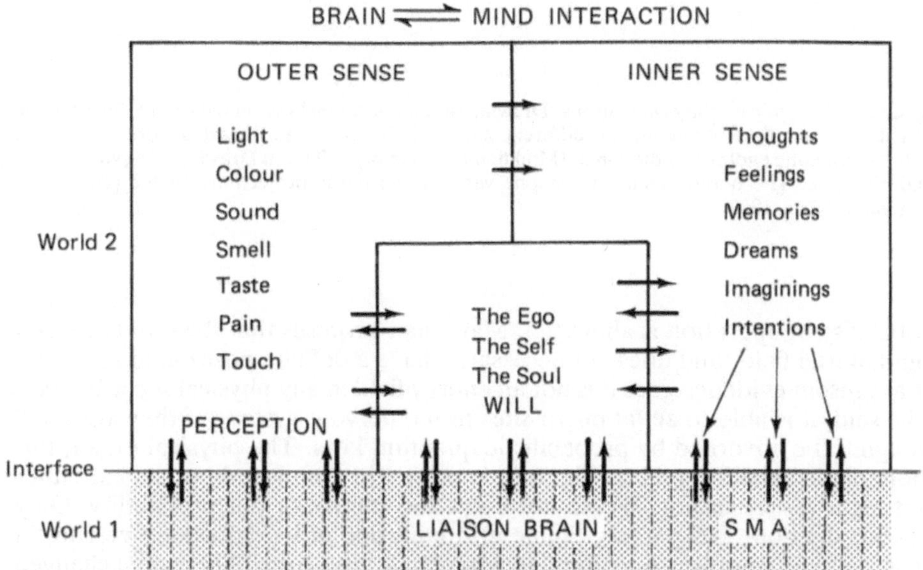

Fig. 6.4. Information flow diagram for brain–mind interaction in human brain. The three components of World 2, namely outer sense, inner sense, and the psyche, self or soul are diagrammed with their communications shown by *arrows*. Also shown are the lines of communication across the interface between World 1 and World 2, that is from the liaison brain to and from these World 2 components. The liaison brain has the columnar arrangement indicated by the vertical broken lines. It must be imagined that the area of the liaison brain is enormous, with open or active modules numbering over a million, not just the two score here depicted.

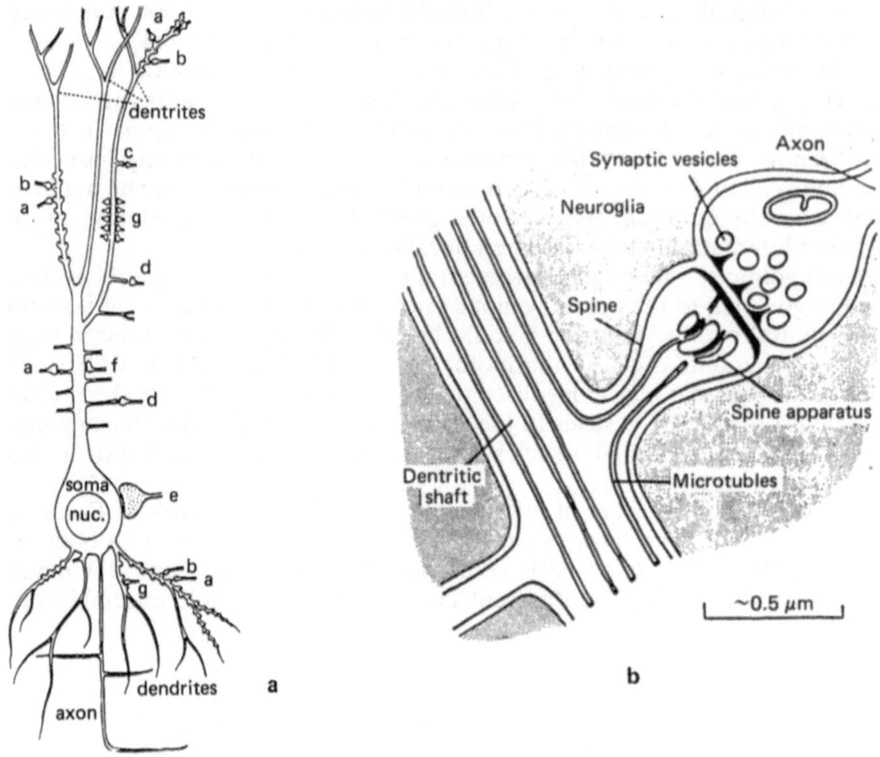

Fig. 6.5. **a** Synaptic endings on neurons. Drawing of a hippocampal pyramidal cell to illustrate the diversity of synaptic endings on the different zones of the apical and basal dendrites, and the inhibitory synaptic endings on the soma. (Modified from Hamlyn 1962). **b** Drawing of a synapse on a dendritic spine. The bouton contains synaptic vesicles and dense projections on the presynaptic membrane. (Gray 1982)

probability of operation is about 0.3. Margenau surmises that the mind acts as a non-material field, and does not necessarily have a definite position in space. So far as present evidence goes it is not an energy field in any physical sense but yet, as he said, it is able to act at micro-sites in the nervous system, if they are small enough to be governed by probabilistic quantum laws. The physical organ, this whole paracrystalline grid in Fig. 6.6, is poised in a multitude of states, either emitting a vesicle or not. We do not know the control of this probability. Only occasionally does this paracrystalline grid emit one of its embedded vesicles in response to an impulse. According to Margenau this probability can be changed by a mental influence acting analogously to a quantum probability field, that carries neither mass nor energy and yet causes changes at micro-sites. He further says that, even if the mind has anything to do with the change, that is, if there is a mind–body interaction, the mind would not be required to furnish energy.

The *first question* that can be raised concerns the magnitude of the effect that could be produced by a probability wave of quantum mechanics. Is the mass of the synaptic vesicle so great that it lies outside the range of the uncertainty

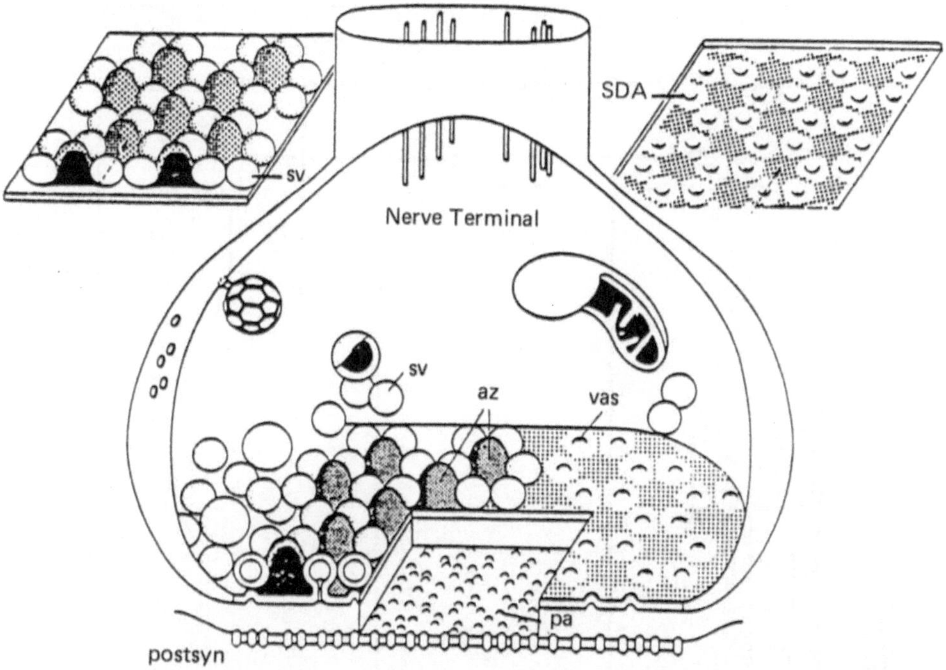

Fig. 6.6. Schema of a mammalian central synapse. The active zone (*az*) is formed by presynaptic dense projections. The postsynaptic aggregation of intramembranous particles is restricted to the area facing the active zone. *sv*, synaptic vesicles; *pa*, particle aggregations on postsynaptic membrane (*postsyn*). Note synaptic vesicles (*sv*) in hexagonal array, as is well seen in the *upper left inset*, and the vesicle attachment sites (*vas*) in the *right inset*. Further description in text. (Akert et al. 1975)

principle of Heisenberg? Margenau (1977, p. 384) adapts the usual uncertainty equation for this calculation of non-atomic situations:

$$\Delta x \cdot \Delta v \geqslant \frac{k}{m} \text{ where } k = 1.06 \times 10^{-27} \text{ erg-s}$$

The mass (m) of a synaptic vesicle 40 nm in diameter can be calculated to be 3×10^{-17}g. If the uncertainty of the position Δx of the vesicle in the presynaptic vesicular grid is taken to be 1 nm, then Δv, the uncertainty of the velocity, comes out as 3.5 nm in 1 ms, which is not far from the right order of magnitude.

I am not saying that the mental influence causes a vesicular emission. I am not saying that the mental influence generates impulses, but merely that it alters the probability of this vesicular emission from the presynaptic vesicular grid, where there are about 50 vesicles lined up (Fig. 6.6), and for some unknown reason only every now and then is one emitted. If we could vary that probability by a mental influence, we would have solved the problem that was bothering Hughlings Jackson. It would be *marvellous* if this could happen! Figure 6.7 shows a cortical pyramidal cell with a few of its 10000 synapses, each with its own grid. So you have to conjecture that the mental influence globally acts on such a nerve cell and alters the probability of emission from many hundreds of its synapses, which would significantly influence the firing of the neuron. Thus you have mental

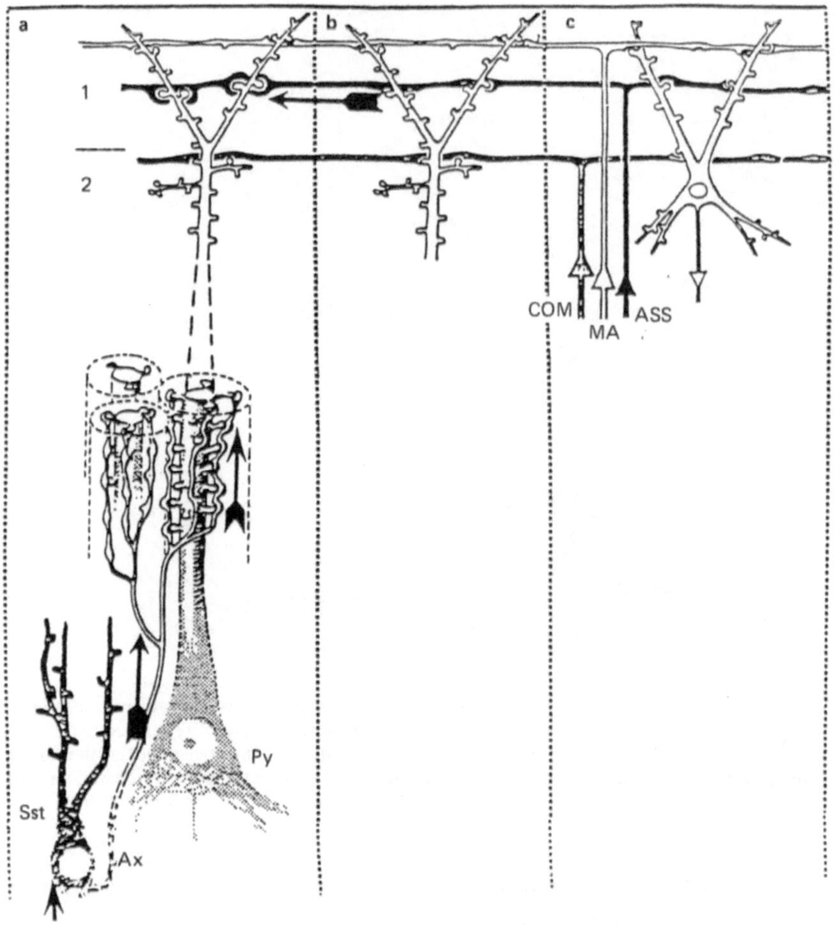

Fig. 6.7a–c. Simplified diagram of connectivities in neocortex showing pathways and synapses. **a–c** show three modules that are vertical functional elements of neocortex, each with about 4000 neurons. In laminae *1* and *2* horizontal fibres arise as bifurcating axons of commissural (*COM*) and association (*ASS*) fibres and also of Martinotti axons (*MA*) from **c**. Horizontal fibres make synapses with apical dendrites of pyramidal cells **a** and **b**. Deeper there is a spiny stellate cell (*Sst*) with axon (*Ax*) making cartridge synapses with shafts of apical dendrites of pyramidal cells (*Py*). (Modified from Szentágothai 1970)

influences working upon the firing of nerve cells without contravening the conservation laws of physics. That is what Jackson would have liked. So the probability of one synapse becomes the reliability of the whole ensemble in a statistical operation. Furthermore, the apical dendrites of cortical pyramidal cells are arranged in clusters, as indicated by the three pyramidal cells partly shown in Fig. 6.7. So the mental influence could act on the whole ensemble of nerve cells that have the same function. Thus we have overcome the two problems Jackson had, and achieved the marvellous outlook he was asking for, and I am sure that he would have been delighted to have seen something that he was looking for and could not explain.

By the same technique as in Figure 6.3 Roland and associates have shown remarkable patches of cerebral activity when the subject was indulging in various set forms of silent thinking. These activated areas were most numerous in the superior zone of the prefrontal cortex, but there were also distinctive patches in widely distributed areas of the association cortex. These findings would be related to Jackson's ideas of the organisation of the cerebral cortex, based on his observations of epilepsy, with different functions in different places. In the so-called silent areas we are now finding increasing evidence of this. So with modern techniques it has been demonstrated that various kinds of mental events do effectively act on the cerebral cortex. This is the kind of conjecture that Popper and I put up some years ago, which we could not account for because we did not go down to the micro-states of the nervous system.

I would like to close on a positive note about Jackson. In introducing a section on subject consciousness, Jackson distinguished between subject consciousness and object consciousness. Subject consciousness is central to World 2, and I quote here, "it would be better to say consciousness of self, better still self consciousness, and best of all, merely self" (Taylor 1931/32, p. 95.4). There in Fig. 6.4 is the self, central to the experiencing being. Jackson continues, "Subject consciousness is something deeper than knowledge; it is that by which knowledge is possible. Perhaps we may say that it is an awareness of our existence as individuals, as persons having the objective states making up for each the (his) universe; it is us in an emphatic sense." That is Jackson's last word on this whole question of the self, and there in Fig. 6.4 it is shown diagrammatically as central to the World of the mind.

References

Akert K, Peper K, Sandri C (1975) Structural organisation of motor end-plate and central synapses. In: Waser PG (ed) Cholinergic mechanisms. Raven, New York, pp 43–57

Gray EG (1982) Rehabilitating the dendrite spine. Trends Neurosci 5:5–6

Hamlyn (1962) An electron microscope study of pyramidal neurones in the Ammon's Horn of the rabbit. J Anat 97:189–201

Margenau H (1977) The nature of physical reality. Oxbow, Woodbridge, Connecticut

Margenau H (1984) The miracle of existence. Oxbow, Woodbridge. Connecticut

Popper KR, Eccles JC (1985) The self and its brain. Springer, Berlin Heidelberg New York

Roland PE, Larsen B, Lassen NA, Skinhoj E (1980) Supplementary motor area and other cortical areas in organisation of voluntary movements in man. J Neurophysiol 43:118–136

Szentágothai J (1970) Les circuits neuronaux de l'écorce cérébrale. Bull Acad R Med Belg VII, X:475–492

Taylor J (ed) (1931/32) Selected writings of John Hughlings Jackson. Vol 2: Evolution and dissolution of the nervous system. Hodder and Stoughton, London. Reprinted (1958) Basic Books, New York

Chapter 7

Hierarchies and Human Memory

A. Baddeley

A perusal of Hughlings Jackson's bibliography reveals that he never wrote a paper specifically on memory. However, the concept of hierarchies has played an interesting and important role in recent theorising about human memory and amnesia.

Consider, first, Jackson's concept of concomitance. The doctrine of concomitance first asserts that states of mind are utterly different from nervous states. Secondly, that the two kinds of state occur together, but for every mental state there is a correlative nervous state. Thirdly, that although states of mind and nervous states occur in parallelism there is no interference of one with the other. Thus, in the case of visual perception, there is an unbroken physical circuit, complete reflex action from sensory periphery through highest centres back to muscular periphery.

The concept of concomitance is, I think, rather similar to the dualist view of mind that was parodied by the philosopher Gilbert Ryle (1949) as the concept of "the ghost within the machine". I think the particular Jacksonian version, based on reflexes, perhaps suggests why theorising about memory was not particularly prominent in Hughlings Jackson's writings.

An analogy can be drawn between the brain and a machine. A reflex concept based very much on physical input and physical output works quite well for a simple physical machine like a piano, or a typewriter. However, with a slightly more complicated machine, like the digital computer, such a view starts to look rather less promising. In the case of memory, the analogy seems even less appropriate. What would be the equivalent of the reflex, and how would one look for the relationship between the physical stimulus and the physical output? Such an approach suggests perhaps an avoidance of complex problems such as aphasia and amnesia, in favour of concentrating on apparently simpler problems such as motor behaviour that might more plausibly be viewed in terms of simple reflexes. I wonder to what extent the aftermath of this avoidance of the complex problem of cognition perhaps still lingers on in British neurology.

To return to hierarchies and memory; I propose to discuss very briefly two approaches to memory and amnesia that have used hierarchical concepts, although not, I must admit, concepts derived directly from Hughlings Jackson. The first of these is an approach to memory that was developed in the early 1970s as a way of describing the relationship between the way in which people process information and how well they remember it. It is the approach developed by Craik and Lockhart (1972), and known usually as the Levels of Processing view of memory. Consider a subject who is expected to remember a word like "CAT"; he could process it purely visually, simply noting that it is made up from upper case letters, printed in black on white, or he could process it more deeply in terms of what it sounds like, that it begins with a C and that it rhymes with "mat". On the other hand, he could process it even more deeply in terms of its meaning, that it is an animal, that it has whiskers and captures mice.

The series of levels through which a word can be processed was conceptualised in terms of perceptual views common at the time, which assumed a hierarchy of perceptual processes. Craik and Lockhart proposed that the further up the hierarchy the word was processed, the more durable the memory trace. A good deal of evidence seemed to support this view, and, as a general rule of thumb, it works reasonably well, although as scientific theory it does have serious limitations (Baddeley 1978).

Cermak (1979) suggested that the concept of Levels of Processing might offer an explanation of amnesia. He proposed that patients might have particular difficulty in remembering because of defective semantic processing leading to a reliance on shallow processing with its associated poor retention. Cermak and his colleagues did indeed find impoverished processing and poor memory in their alcoholic Korsakoff patients (Cermak 1979). Unfortunately, it has subsequently become clear that this result is attributable to the presence of general cognitive impairment in the patients (Cermak 1982). Other patients who are not amnesic may show similar processing deficits (Moscovich 1982), while patients may be densely amnesic in the absence of such cognitive deficits (Baddeley 1982).

A number of studies have tackled the Levels of Processing approach to amnesia by studying the effect on learning of ensuring that the amnesic patient processes material deeply. I will quote just one study, which I think might particularly appeal to Hughlings Jackson, who wrote a paper on the psychology of jokes. Meudell et al. (1980) showed their patients cartoons and set tasks requiring either a very low level of analysis in which the subject looked and tried to decide what differences, if any, there were between two virtually equivalent cartoons, or else a deeper analysis, in which the patient had to describe what was going on in the cartoons. A third task involved a simple instruction to try to learn the cartoons. The results showed that although the alcoholic Korsakoff patients had poorer memory than the controls, they showed the same effect of depth of encoding. There was no evidence therefore for either a spontaneous unwillingness to encode deeply, or a failure to benefit from such coding.

There are a number of other reasons for suspecting that the levels approach is less than ideal. For example, the initial hierarchical view of perception on which it was based was subsequently proved to be oversimplified; it is now clear that perception does not comprise a simple linear hierarchy of stages, but involves interaction between stages, involving processes interacting in a heterarchy rather than a hierarchy.

There are also problems of how to define and measure depth of processing, as well as demonstrations that, in some circumstances, shallow processing can lead to very durable long-term memory. In conclusion then, while it remains a good rule of thumb, the concept of Levels of Processing leaves a good deal to be desired as a scientific theory.

I shall now discuss another more recent use of the concept of hierarchy in memory. It is an approach that developed from observations of memory in amnesic patients and is now being incorporated into theories of normal memory. It is concerned with the question of what aspects of memory are impaired and what aspects are intact in classic global amnesic patients, for example patients suffering from Korsakoff's syndrome.

Such patients exhibit many everyday memory problems. They cannot remember what they had for breakfast, they do not recognise you if you have met them before, cannot find their way around and tend not to be well orientated in time or space, often not knowing where they are, what day it is or how old they are. In terms of psychological tests they perform badly on a very wide range of tasks, whether they involve recalling words or stories, or recognising pictures or faces (Baddeley 1982). On the other hand, their immediate memory for small amounts of material is normal, but once you exceed their immediate memory span of about six numbers or words, performance deteriorates rapidly. The same pattern occurs whether the patient is recalling verbal material like numbers, or visual sequences of lights. If you present the subject with ten unrelated words and ask him to recall them immediately, then the last few, which appear to be recalled from some temporary short-term store, will be recalled as well by the amnesic patient as by controls, while performance on earlier items is much poorer than controls, as is clear from Fig. 7.1.

The evidence therefore seemed to suggest normal performance by amnesic patients when recall is based on a temporary short-term or working memory system, coupled with grossly impaired performance on tasks demanding more durable or long-term memory (Baddeley and Warrington 1970).

In recent years, however, there has been growing evidence that certain types of long-term learning are preserved, even in densely amnesic patients. There is, for example, the classic story of the occasion when Claparède (1911) secreted a pin in

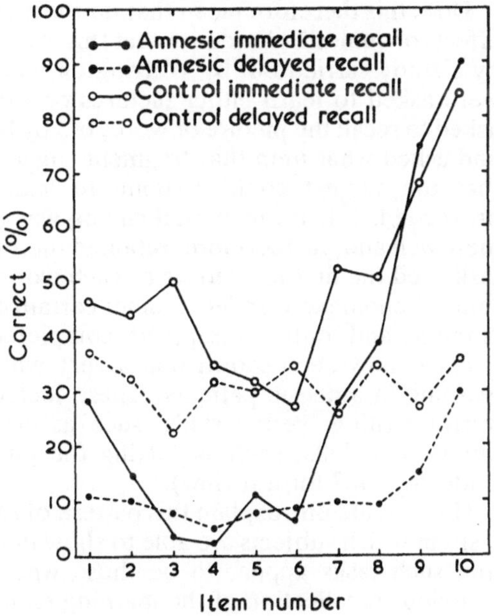

Fig. 7.1 Performance of amnesic and control patients on a task involving the immediate and delayed recall of lists of unrelated words. Amnesic patients are impaired on all except the immediate recall of the last few words presented on any given list, the so-called recency effect. (Data from Baddeley and Warrington 1970)

his hand while shaking hands with a Korsakoff patient. Next morning when he offered to shake hands, the patient was reluctant to do so, but could not report or recollect why.

This remained a relatively isolated observation until the 1960s when other examples of apparently normal learning started to emerge. For example, casual conversation with amnesic patients suggests that they can use language perfectly normally. In a more formal test of their ability to process language, Barbara Wilson and I asked two pure amnesic patients to make semantic judgements. The task required them to decide whether pairs of words were both from the same category or not; for instance, "leopard, dog" should evoke a "yes" response while "cabbage, table" should not. Our amnesic patients took about 1.5 s per decision, just the same as our normal control subjects, while their error rate for this task was, if anything, slightly less. Similarly, when asked to judge the truth of sentences about the world, such as "Robins have red breasts", or "Robins hold political office", the amnesic patients responded just as rapidly as normal subjects and with just as low an error rate. Hence, there is at least one aspect of memory that appears to be unimpaired; this type of memory reflects knowledge of the world and is often referred to as *semantic memory*, in contrast to *episodic memory*, a term that refers to memory for specific personally experienced events. Episodic memory is of course grossly impaired in amnesic patients.

A third aspect of memory that is unimpaired in amnesia is essentially an extension of the Claparède phenomenon described earlier. A number of studies have noted that amnesic patients appear to show relatively normal motor learning (e.g. Corkin 1968). A study by Brooks and myself required amnesic patients to learn a tracking task in which the patient tries to keep a stylus in contact with a rotating spot of light; performance is measured in terms of how long the subject can keep on target over a standard period of time (Brooks and Baddeley 1976). We found excellent and exactly equivalent learning in our amnesic and our control subjects.

Does this therefore mean that motor learning is specially protected against the effects of amnesia? Evidence that this is almost certainly not the case is suggested by a study carried out by Warrington and Weiskrantz (1968) in which subjects were asked to learn either pictures or words. They were tested, not by being asked to recall the picture or word, but by being shown a fragment of the original, and asked what item that fragment suggested. Learning was shown by the fact that the subject could respond to lesser and lesser fragments as learning proceeded. When this partial cueing was used, amnesic patients learned extremely well and, furthermore, retained their learning. A subsequent version of this task used the first few letters of each word as a test cue and showed exactly the same phenomenon. Indeed, under certain circumstances, the difference between amnesics and controls disappears completely (see Shimamura 1986 for a review).

It has now been shown that a very wide range of tasks will show preserved learning in amnesic patients. These include perceptual tasks, such as reading mirror writing; verbal skills, such as are involved in solving anagrams; and intellectual skills, such as solving the puzzle called the Tower of Hanoi (see Baddeley 1982 for a review).

How should one explain this pattern of results? What seems to characterise the tasks in which subjects are able to show normal learning despite being amnesic is that such tasks appear to be those where performance does not depend on conscious recollection of the learning experience. Characteristically an amnesic

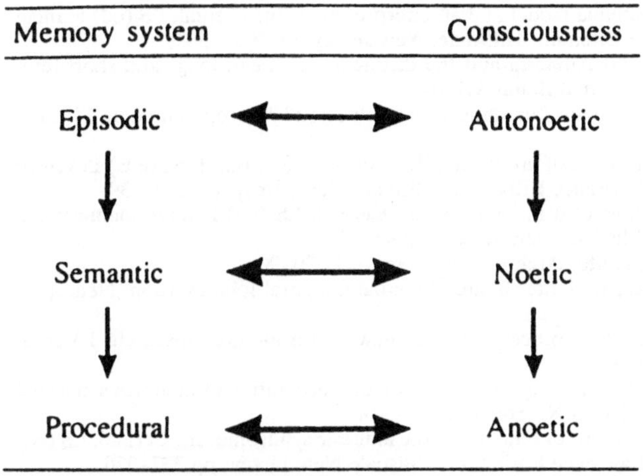

Fig. 7.2. Tulving's hierarchical conceptualisation between varieties of memory and the degree of conscious awareness involved at each level. (Based on Tulving 1985)

patient will often show normal learning over trials at the same time as he denies ever having seen the task before. However, such tasks can typically be performed without any need to recollect earlier practice trials. Such skill acquisition is often termed *procedural learning*.

One way of conceptualising this overall pattern of results is that suggested by Tulving (1985), who postulates a hierarchy of different types of memory, as shown in Fig. 7.2.

Procedural learning involving the acquisition of some form of intellectual, perceptual or verbal skill is assumed to form the lowest level of the hierarchy. Semantic memory, which represents knowledge of the world, constitutes the intermediate level while episodic memory is at the highest level, Tulving assumes that these three levels of learning are associated with three levels of consciousness. Episodic memory is assumed by Tulving to be *autonoetic*, by which he means that the subject is aware of the process of remembering. Semantic memory is assumed to involve a rather lower level of conscious awareness, and is termed noetic. Tulving assumes that procedural learning does not require any conscious awareness, a state he refers to as *anoetic*.

I suspect that Hughlings Jackson would have found Tulving's views rather attractive and consistent with his own view that "Less consciousness attends activities the more organised and automatic they are, or have become, the highest least organised, least automatic, most imperfectly reflex centres are the physical basis of the most vivid consciousness." So although he did not talk about memory, I think Jackson would have recognised and sympathised with at least some of our more hierarchical current views.

References

Baddeley AD (1978) The trouble with levels: a re-examination of Craik and Lockhart's framework for memory research. Psychol Rev 85:139–152

Baddeley AD (1982) Amnesia: a minimal model and an interpretation. In: Cermak LS (ed) Human memory and amnesia. Lawrence Erlbaum, Hillsdale, New Jersey, pp 305–336
Baddeley AD, Warrington EK (1970) Amnesia and the distinction between long- and short-term memory. J Verbal Learning and Verbal Behav 9:176–189
Brooks DN and Baddeley AD (1976) What can amnesic patients learn? Neuropsychologia 14: 111–122
Cermak LS (1979) Amnesic patients' level of processing. In: Cermak LS, Craik FIM (eds) Levels of processing in human memory. Lawrence Erlbaum, Hillsdale, New Jersey, pp 119–139
Cermak LS (1982) The long and short of it in amnesia. In: Cermak LS (ed) Human memory and amnesia. Lawrence Erlbaum, Hillsdale, New Jersey, pp 43–60
Claparède E (1911) Récognition et moite. Arch Psychol Geneve 11:79–90
Corkin S (1968) Acquisition of motor skill after bilateral medial temporal-lobe excision. Neuropsychologia 6:255
Craik FIM, Lockhart RS (1972) Levels of processing: a framework for memory research. J Verbal Learning Verbal Behav 11:671–684
Meudell P, Mayes A, Neary D (1980) Orienting task effects on the recognition of humorous material in amnesia and normal subjects. J Clin Neuropsychol 2:1–14
Moscovich M (1982) Multiple dissociations of function in the amnesic syndrome. In: Cermak LS (ed) Human memory and amnesia. Lawrence Earlbaum, Hillsdale New Jersey, pp 337–370
Ryle G (1949) The concept of mind. Hutchinson, London
Shimamura A (1986) Priming effects in amnesia: evidence for a dissociable memory function. Q J Exp Psychol 38A:619–644
Tulving E (1985) Memory and consciousness. Can Psychol 26:1–12
Warrington EK, Weiskrantz L (1968) A new method of testing long-term retention with special reference to amnesic patients. Nature 217:972–974

Section III

EPILEPSY

Chapter 8

Hughlings Jackson and Epilepsy: An Introduction

E. H. Reynolds

Although some of Jackson's ideas on hierarchies have come in for some muted criticism, I can state with confidence that his observations on epilepsy have endured and will continue to endure (Reynolds 1988). Collectively they rank amongst the greatest contributions to our knowledge and understanding of epilepsy, perhaps on a par with the Hippocratic writings on the Sacred Disease, with the exception that although the latter were written in Greek they were much easier to read! Of the two volumes of John Hughlings Jackson's selected writings edited by Taylor 1931/32, one volume is devoted wholly to epilepsy and epileptiform convulsions. This is a true reflection of his interest and contributions. He wrote more about epilepsy than any other topic, including hierarchies, evolution and dissolution. He wrote regularly over a 40-year period from the time he began work at the National Hospital for the Paralysed and Epileptic in 1862.

Epilepsy was for Jackson not only a worthy study in its own right, but also because it was a model for the study of so many other neurological problems; and also for the insight it gave into the functioning of the nervous system. His were the most extensive and definitive studies of unilateral convulsions (Jackson 1869), and this is reflected in our use today of the term "Jacksonian epilepsy". He derived great pleasure from the fact that some of his ideas were confirmed by the experimental studies of Fritsch and Hitzig, and also Ferrier. They acknowledged their great debt to him, and he to them.

As this book is concerned with hierarchies, I will refer to his hierarchical concepts of epilepsy, best described in his Lumleian Lectures (Jackson 1890). Lowest level fits he described as "pontobulbar". Middle level fits, associated with the motor cortex and corpus striatum – the unilateral seizures which today we call "focal" or "partial" – he referred to as "epileptiform". Highest level fits, which today we might call "generalised" or "idiopathic", he described as "epilepsy proper". Jackson was not so much concerned with the perennial issue of the nosological classification of clinical seizures, but he was very precise in the terms he used, such as "epileptiform" and "epileptic". Those readers who are members of the International League Against Epilepsy will know that from time to time the League convulses itself by attempting yet another nosological classification of seizures or epilepsy. Jackson was more concerned with other questions. At what level in the nervous system does the seizure arise? What is the state of the nervous tissue during a seizure, in this experiment of nature, in this departure from health? Although there were humoral and vascular theories before Jackson's time, he proposed and developed the first neuronal theory of epilepsy, which is the foundation of our modern scientific understanding of epilepsy. This is epitomised in his famous physiological definition of epilepsy (Jackson 1873): "Epilepsy is the name for occasional, sudden, excessive, rapid and local discharges of grey matter".

He presented this view, he represented it, and re-represented it, in a series of articles over a long period of time. If repetition ever invests a principle with truth, then this would be a very good example of it. I have one slight reservation about this intuitive statement. I do wish he had used another word instead of "epilepsy" in this definition, because it opened up a Pandora's box, as he soon began to realise. The problem is illustrated in the following statement (Jackson 1873): 'A sneeze is a sort of healthy epilepsy".

Any clinical problem which is vaguely paroxysmal is vulnerable to being called epilepsy. Epilepsy is not only a clinical diagnosis, but it is now the property of neurophysiologists, who describe epileptic or epileptiform electroencephalograms, or even epileptic neurons. I will not dwell on this semantic issue now, but neurologists faced with the problem of relating clinical events to EEG findings will know exactly what I mean, and this problem can be traced back to Jackson's difficulties in communication.

References

Jackson JH (1869) A study of convulsions. Trans St. Andrews Med Grad Assoc iii:162–204

Jackson JH (1873) On the anatomical, physiological and pathological investigation of epilepsies. West Riding Lunatic Asylum Med Rep iii:315–339

Jackson JH (1890) On convulsive seizures. Br Med J I:703–707, 765–771, 821–827

Reynolds EH (1988) Hughlings Jackson: a Yorkshireman's contribution to epilepsy. Arch Neurol 45:675–678

Taylor J (ed.) (1931/32) Selected Writings of John Hughlings Jackson, vols 1 and 2. Hodder and Stoughton, London. Reprinted (1958) Basic Books, New York

Chapter 9

Psychiatric Aspects of Temporal Lobe Epilepsy

M. R. Trimble

The relationship between epilepsy and psychiatry has been of interest since the earliest medical writings. Hippocrates and Aretaeus both commented on associations between behavioural disturbances and the condition we now refer to as epilepsy, and nearer our present time the psychiatrists of the eighteenth and nineteenth centuries from several European countries observed psychopathology in epileptic patients, mainly in an institutionalised setting. While the modern era of epilepsy and epileptic research may be said to stem from the introduction of the electroencephalogram after the Second World War, many of the ideas that were then revived can be directly related back to the works of Hughlings Jackson.

Other authors in this book have discussed Jackson's great contributions to the understanding of epilepsy, one of them being the clarification of the concept of that form of epilepsy which provokes, in the main, partial seizures. Prior to the writings of Jackson, the term "epilepsy" tended to suggest a generalised convulsive disorder, which affected mainly motor function. Although sensory phenomena had been described, and the aura was well known, the concept that only part of the brain could be involved in the expression of an epileptic seizure had only been hinted at, for example in the writings of Pritchard (1822), Todd (1856) and others. Jackson elaborated more fully on such partial episodes, a motor form being ultimately designated "Jacksonian" by Charcot. Of particular relevance for the psychiatric aspects of epilepsy was Jackson's work on "the uncinate group of fits", and patients who presented with "dreamy states". He thus clearly defined how alteration of the cognitive or mental state could be found in association with a seizure.

These ideas were rejuvenated with the introduction of the electroencephalogram and the clear delineation of temporal lobe epilepsy and associated psychomotor seizures. In our present classifications, the terms "simple" and "complex partial seizures" are used to describe the kind of attacks which Jackson so clearly delineated.

Jackson himself seems to have been aware of the relationship between epilepsy and insanity. It is known that some of his earliest teachers in medicine were psychiatrists, and in his writings there are frequent references to insanity. He quoted with approval the statistic given by Tuke, namely that 6% of cases of insanity in asylums were caused by epilepsy (Jackson 1875). In considering the changes in thinking which have occurred in relationship to the theme of epilepsy and psychiatry, the analysis of Guerrant et al. (1962) proves useful. These authors identified four periods of change, as shown in Table 9.1. In the latter part of the nineteenth century (period of epileptic deterioration) it was considered that the very fact that patients had epilepsy was enough to assume some form of deterioration of personality and intellect. This view stemmed more directly from the degeneracy theory of mental disorder so prevalent in continental Europe in the nineteenth century, in which it was considered that psychopathology, including epilepsy, was the result of a progressive hereditary degenerative condition, which could pass from one generation to the next, becoming more severe with succeeding progeny.

Table 9.1. Changing ideas of the relationship between epilepsy and psychopathology[a]

Period of epileptic deterioration	(–1900)
Period of the epileptic character	(1900–1930)
Period of normality	(1930–)
Period of psychomotor peculiarity	(1930–)

[a] After Guerrant et al. (1962).

In the earlier part of this century, however, a change of thinking occurred (period of the epileptic character). The epilepsy itself and associated behavioural or personality changes became seen as secondary to some underlying third principle, which was common to both. One of the uniting factors was the personality structure, the constitution of the patient – a line of thought persistently pursued by authors such as Pierce-Clark (1929). It was claimed that it was possible to discern certain patterns of developmental personality traits which could be identified even prior to the onset of seizures, and which implied an epileptic constitution. This whole era became bound up with Freudian psychodynamics, such that epilepsy, as with the neuroses, was thought to reflect on infantile motives and inadequate development of individual affects and instincts leading to the later development of symptoms. From these ideas arose concepts of the "epileptoid" character and of epileptic equivalents, which then became confused with psychomotor epilepsy and the idea that personality change could occur secondary to temporal lobe lesions. Thus, the "period of psychomotor peculiarity" is still with us today. No longer is the personality of patients with epilepsy seen as a reflection of some form of constitutional deficit; instead, the effects of a chronic temporal lobe lesion on behaviour and personality have been investigated, and the concept that change of personality in temporal lobe epilepsy may reflect an organic brain syndrome has been pursued.

An alternative view, which is upheld in many centres, is that there is nothing special about temporal lobe epilepsy, and that all patients with epilepsy, if allowed to, will have normal personality development and not be liable to undergo any change in their mental state or personality profile. It is acknowledged that cerebral damage may occur secondarily to recurrent seizures with head injuries or bouts of anoxia and that personality change may occur as a consequence of social stigmatisation. This alternative view, however, allows for none of the biological determinism inherent in the concept that a temporal lobe lesion may of itself provoke and produce psychopathology.

In reviewing the literature on epilepsy and psychiatry, it is most interesting to consider personality disorders and psychoses in relationship to epilepsy, and to be concerned primarily with inter-ictal psychiatric conditions. Thus, most physicians would agree that a patient's mental state may be temporarily deranged by the electrical disturbance that accompanies an ictus, and the presentation of cognitive disruption with an acute psychosis (an acute organic brain syndrome or delirium) is not infrequently seen on medical wards where patients with epilepsy are admitted. The controversy, however, relates to chronic behaviour problems, and their relationship to the epileptic disturbance within the brain.

With regard to personality disorder, one problem has been that attempts using standardised rating scales of personality to define differences between those with temporal lobe and other forms of epilepsy have often proved negative (Trimble and Perez 1980). Although these studies tend to demonstrate that patients with epilepsy, when compared with other patients who do not have epilepsy, show significantly more psychopathology, the interpretation of this is compatible with either of the above views and the centre of the argument relates to differences between patients with temporal lobe epilepsy and others. Some recent studies clarify this issue considerably. Bear and Fedio (1977) developed their own rating scale of 18 behavioural features, drawn from the literature, which were supposedly associated with temporal lobe epilepsy. They used this scale with a group of patients with temporal lobe epilepsy and compared their profiles with controls with neuromuscular disease and with healthy counterparts. Their scale appeared to separate out the temporal lobe group, in particular with regard to such traits as humourless sobriety, dependency, obsessionality, and religious and philosophical concerns. Interestingly, they noted differences in the personality profiles of patients with left-, when compared with right-sided temporal lobe foci. The former scored more for anger, paranoia and dependence, whereas the latter had higher ratings for elation. Although these data have been criticised, they have been partially replicated by other groups (Hermann and Reil 1981). In another study Nielsen and Kristensen (1981) used the same scale in patients with either a mediobasal temporal lobe EEG focus or a lateral site of their abnormality. The former had significantly more hypergraphia, elation, guilt and paranoia in comparison with those with a lateral focus, and again patients with left-sided abnormalities scored higher than those with changes on the right. Using the Minnesota Multiphasic Personality Inventory (MMPI), Hermann et al. (1982) examined patients with temporal lobe epilepsy, separating them out into different groups dependent upon their auras. Those with an aura of fear scored significantly higher on several items of psychopathology than those without such auras or patients with generalised epilepsy. In particular, those with auras of fear scored high on assessments for psychosis.

These recent data emphasise two important points. First, attempts to examine personality changes in epilepsy by purely clinical means may provide only one avenue of investigation; the use of quantified and validated rating scales may provide a more accurate and different kind of assessment. Secondly, temporal lobe epilepsy does not represent a uniform condition. These findings suggest that a group of patients with mediobasal lesions, and possibly with an aura of fear, which suggests peri-amygdaloid discharges, are more prone to psychopathology. The medially placed, limbic system lesions are here contrasted with the more laterally placed neocortical temporal lobe abnormalities, and the suggestion is that patients with chronic limbic system epilepsy may be more susceptible to psychopathology than others.

The concept of personality change in relation to temporal lobe lesions has been most strongly made by Geschwind and his colleagues (Waxman and Geschwind 1975; Geschwind 1979). They emphasise in particular hyposexuality, religiosity, and hypergraphia – a tendency towards extensive and compulsive writing. The hypergraphia is particularly interesting (Trimble 1986). It is characteristically meticulous, and moral or religious overtones are often noted in the content. There is preoccupation with detail, and often a compulsive quality to much that is written. Repetition of words or sentences may be seen, and variants include the hiring of public stenographers by patients or extensive drawing or painting. Geschwind felt this was a disregarded neurological sign of temporal lobe dysfunction and one which was often apparent if sought for when examining patients.

If the temporal lobes thus appear to be related to personality organisation and pathology of the temporal lobes relates to changes of behaviour, it is of further interest that a relationship has also been described between temporal lobe epilepsy and inter-ictal psychosis. It was Slater (Slater and Beard 1963) who most clearly drew attention to a schizophrenia-like psychosis of epilepsy, which occurred in patients with epilepsy with increased frequency. In particular, he noted the absence of abnormal schizoid premorbid personality traits or a family history of psychiatric disturbance in these patients that might suggest a predisposition to schizophrenia. The mean age of onset of the psychosis was 29.8 years, occurring after the epilepsy had been present for a mean of 14.1 years. Although in most patients it was not possible to relate the onset of the mental illness to any change in the quantity or quality of seizures, in some 25% of cases the psychotic symptoms appeared as the frequency of generalised seizures was falling. While neurological findings were generally negative, air-encephalography showed abnormalities in the majority of patients, often involving dilation of one or both temporal horns. In recent years, several other authors have investigated the inter-ictal psychoses of epilepsy, and in general they appear to be associated with medial basal EEG discharges, the presentation clinically with automatisms (Kristensen and Sindrup 1978) and, if schizophrenia-like, a disturbance mainly of the dominant temporal lobe (Flor-Henry 1969; Trimble 1983; Perez et al. 1985). Since the phenomenology of such patients closely resembles that of schizophrenia in the absence of epilepsy, it is suggested that temporal lobe epilepsy is a good model for development of psychoses in some patients, that is, for seeking the neurological underpinning of psychotic phenomena, such as hallucinations, delusions or paranoia.

It is here that we again recall the extensive writings of Hughlings Jackson on the relationship between cerebral lesions and symptomatology. His introduction of the concepts of positive and negative symptoms has recently been revived in neuropsychiatry, but mainly in relation to schizophrenia. Current discussion has none of the subtlety of the earlier Jacksonian ideas and divides positive and negative symptoms on purely clinical grounds rather than on a basis of trying to understand cerebral function in a dynamic context (Trimble 1986). The psychosis of temporal lobe epilepsy thus seems to be a model mainly for the positive symptoms of psychosis (hallucinations especially), similar to those seen in schizophrenia without epilepsy, and suggests a role of the temporal lobes in their pathogenesis. The inter-ictal disturbances in the temporal lobes of epileptic patients which we are able to identify with modern technology such as the PET scan, which shows continuous hypometabolism and widespread low blood flow in the temporal lobes and beyond (Bernardi et al. 1983), may directly relate to some symptoms, e.g. memory disturbances of epileptic patients, but are unlikely to provoke these positive symptoms directly. The latter surely relate to the activity of surrounding, less affected neocortical structures, whose interplay with the limbic system in patients with temporal lobe lesions is so markedly disturbed. The emphasis on the dominant cortex may have as much to teach us regarding the role of this hemisphere for symbolic thought and its disorganisation as the earlier observations of Broca had for developing our ideas of language and its localisation.

References

Bear DM, Fedio P (1977) Quantitative analysis of inter-ictal behaviour in temporal lobe epilepsy. Arch Neurol 34:454–467

Bernardi S, Trimble MR, Frackowiak RSJ, Wise RJS, Jones T (1983) An inter-ictal study of partial epilepsy using positive emission tomography and the oxygen-15 inhalation technique. J Neurol Neurosurg Psychiatry 46:573–477

Flor-Henry P (1969) Psychosis and temporal lobe epilepsy. Epilepsia 10:363–395

Geschwind N (1979) Behavioural changes in temporal lobe epilepsy. Psychol Med 9:217–219

Guerrant J, Andersen C, Fischer A, Weinstein MR, Jarros RM, Deskins A (1962) Personality and epilepsy. Thomas, Springfield, Illinois

Hermann B, Reil A (1981) Inter-ictal personality and behavioural traits in temporal lobe and generalised epilepsy. Cortex 17:125–128

Hermann BP, Dickman S, Schwarz MS, Karnes WE (1982) Inter-ictal psychopathology in patients with ictal fear: a quantitative investigation. Neurology 32:7–11

Jackson JH (1875) On temporary mental disorders after epileptic paroxysms. West Riding Lunatic Asylum Med Rep V:105

Kristensen O, Sindrup EH (1978) Psychomotor epilepsy and psychosis. Acta Neurol Scand 57:370–379

Nielsen H, Kristensen O (1981) Personality correlates of sphenoidal EEG foci in temporal lobe epilepsy. Acta Neurol Scand 64:289–300

Perez MM, Trimble MR, Reider I, Murray NM (1985) Epileptic psychosis, a further evaluation of PSE profiles. Br J Psychiatry 146:155–163

Pierce-Clark L (1929) A psychological interpretation of essential epilepsy. Brain 43:38–49

Pritchard JC (1822) A treatise on diseases of the nervous system. Thomas and George Underwood, London

Slater E, Beard AW (1963) The schizophrenia-like psychoses of epilepsy. Br J Psychiatry 109:95–150

Todd RD (1856) Clinical lectures on paralysis, certain diseases of the brain, and other affections of the nervous system. London (publisher unknown)

Trimble MR (1983) Inter-ictal behaviour and temporal lobe epilepsy. In: Pedley T, Meldrum B (eds) Recent advances in epilepsy, vol 1. Churchill Livingstone, Edinburgh, pp 211–229

Trimble MR (1986) Positive and negative symptoms in psychiatry. Br J Psychiatry 148:587–589

Trimble MR, Perez MM (1982) The phenomenology of the chronic psychoses of epilepsy. In: Koella WP, Trimble MR (eds) Temporal lobe epilepsy, mania, schizophrenia and the limbic system. Karger, Basel, pp 98–105

Waxman SG, Geschwind N (1975) The inter-ictal behaviour syndrome of temporal lobe epilepsy. Arch J Psychiatry 32:1580–1586

Chapter 10

Photosensitive Epilepsy and Visual Discomfort

A. Wilkins

Photosensitive epilepsy is interesting partly because it is the most common form of reflex epilepsy, and the discovery of techniques for preventing seizures may be of practical significance, and partly because the visual system is better understood than other sensory systems, and inferences about physiological mechanisms can therefore be made. As will be shown in this chapter, the inferences may help explain not only seizures but also the visual discomfort experienced by people who do not have epilepsy.

The chapter is divided into three sections, the first concerning photosensitive epilepsy (what stimuli provoke seizures, where, when and how they are initiated), the second concerning visual discomfort and its similarity to photosensitive epilepsy, and the third, techniques for preventing these disorders.

Photosensitive Epilepsy

What Visual Stimuli Are Responsible for Seizures?

A wide variety of visual stimuli can produce attacks, ranging from diffuse flicker, to stationary, steadily illuminated patterns. A large number of patients are sensitive to patterns as well as to flicker, but only when the patterns have certain very specific spatial characteristics. When photosensitive patients look at patterns with these characteristics the EEG will often exhibit epileptiform, paroxysmal EEG abnormalities. The occurrence of these abnormalities is strictly dependent on the nature of the pattern stimulation. The patterns that do and do not induce the response give us important clues as to where the seizures are triggered in the visual system.

Where Are the Seizures Triggered?

The seizures appear to be triggered in the visual cortex. There are four reasons:

1. The probability of an epileptiform response is dependent on the length of line contour in the pattern: the longer the line contour the greater the probability of paroxysmal abnormalities (Wilkins et al. 1979a). We know that in the visual cortex of the brain, neurons are sensitive to lines (e.g. Hubel and Wiesel 1979).

2. When one eye sees a pattern of horizontal stripes and the other eye a pattern of vertical stripes, epileptiform EEG activity is less likely than when both retinae receive the same pattern orientation (Wilkins et al. 1979a). The seizure mechanism evidently involves some degree of interaction between the eyes, and it is at the level of the visual cortex that a substantial proportion of neurons are binocular (e.g. Hubel and Wiesel 1979).

3. In certain patients including those without any evidence of astigmatism, sensitivity is confined to a restricted range of pattern orientations (Wilkins et al. 1980). Cortical neurons are orientation selective (e.g. Hubel and Wiesel 1979).

4. The spatial characteristics of stimulation for which paroxysmal activity is most likely can be independent of the temporal characteristics. For example, if a pattern of stripes is vibrated in the direction orthogonal to the stripes, the spatial frequency (line width) of the patterns that are most likely to induce paroxysmal abnormalities is independent of the temporal frequency of vibration (Fig. 10.1).

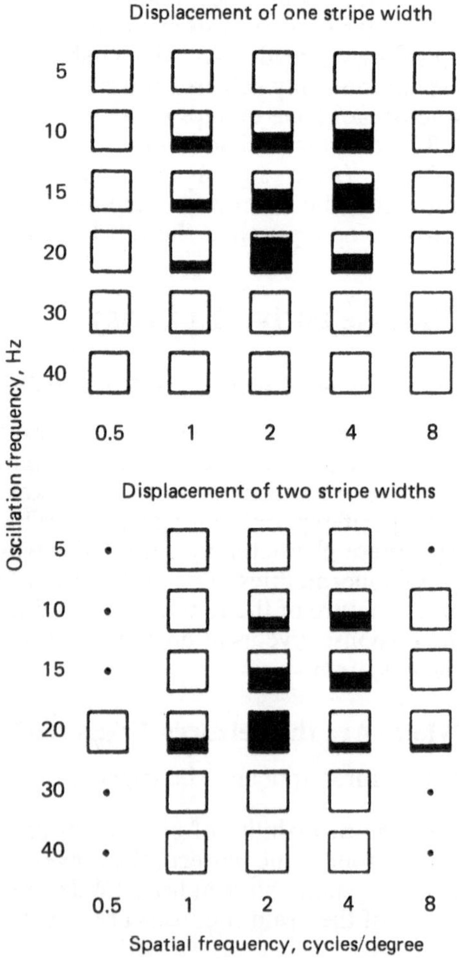

Fig. 10.1. Probability of paroxysmal EEG activity as a function of the spatial frequency, temporal frequency and amplitude of movement of a vibrating pattern of stripes. The height of the *shaded area* within each box is proportional to the probability of paroxysmal activity averaged across patients. The pattern had a square-wave luminance profile. The amplitude of movement was equal to the width of one stripe (half a spatial cycle) or two stripes (one complete spatial cycle of the pattern) and the results are tabulated separately, as shown. The movement was in a direction orthogonal to the stripes, and the profile of displacement against time was triangular.

Cells in the visual cortex that respond to a particular width of a bar do so over a range of positions, and so the independence of spatial and temporal characteristics is consistent with induction in the cortex (Binnie et al. 1979). Unfortunately, these inferences are complicated by the interaction between space and time that occurs in the level of the retina (e.g. Masland 1986).

5. The final line of evidence to support a cortical locus comes not from the characteristics of the patterns, but the topography of the epileptiform activity. If a pattern appears in the left visual field then paroxysmal abnormalities are seen over the right posterior quadrant, and vice versa. In some patients patterns in the two upper quadrants will induce paroxysmal abnormalities with a lower topography on the scalp than patterns in the lower two visual quadrants. As illustrated in Fig. 10.2, the scalp topography of the epileptiform activity follows that of the underlying visual cortex (Wilkins et al. 1981; Darby et al. 1986).

When Do the Seizures Occur?

The answer to the above question is complex, and concerns (1) the extent of excitation, (2) the interaction of the cerebral hemispheres and (3) the role of synchronisation.

Seizures appear to arise when a normal physiological excitation within one hemisphere exceeds a critical mass. We know that the physiological excitation is normal because photosensitive patients can have normal vision inter-ictally, normal acuity, normal stereopsis, and indeed normal visual sensitivity to patterns at a very low contrast, even those that, at higher contrasts, would induce paroxysmal EEG abnormalities. The evidence that a critical mass of excitation is necessary for epileptogenesis comes from the effects of pattern size. The size of the pattern is perhaps the most important characteristic determining whether or not paroxysmal activity will occur. The size of the pattern can be manipulated in many different ways, for example by varying the outer radius of a centrally fixated pattern, (compare a and b in Fig. 10.3), or the inner radius (b and c in Fig. 10.3), or by varying the number and angular size of a pattern of rings, cut to resemble the slices of a cake (d, e and f in Fig. 10.3). The area of the cortex to which the pattern projects determines the epileptogenic properties of the pattern. The small disc, such as pattern a in Fig. 10.3, may project to the same cortical area as a relatively large annulus such as pattern c in Fig. 10.3. A pattern of two slices projects to a similar area of cortex as a pattern of four slices with half the angular size. When the cortical representation is approximately equated, all the various types of pattern have equivalent effects (Wilkins et al. 1980), although, in general, patterns with curved segments are less epileptogenic than those in which the segments are linear. The seizure activity appears to be triggered when normal physiological excitation within any region of the visual cortex exceeds some critical level. Within either hemisphere the visual cortex may be equipotential for seizure activity.

The patterns a–f in Fig. 10.3 are symmetrical either side of the central fixation point. Both cerebral hemispheres therefore receive equivalent stimulation via the lateral visual fields. If the stimulation is not symmetrical but asymmetric, as in patterns g and h in Fig. 10.3, very different effects are seen. First, paroxysmal abnormalities are far more likely to occur in response to a pattern in one lateral visual field than in response to a pattern of equivalent size in the upper or lower

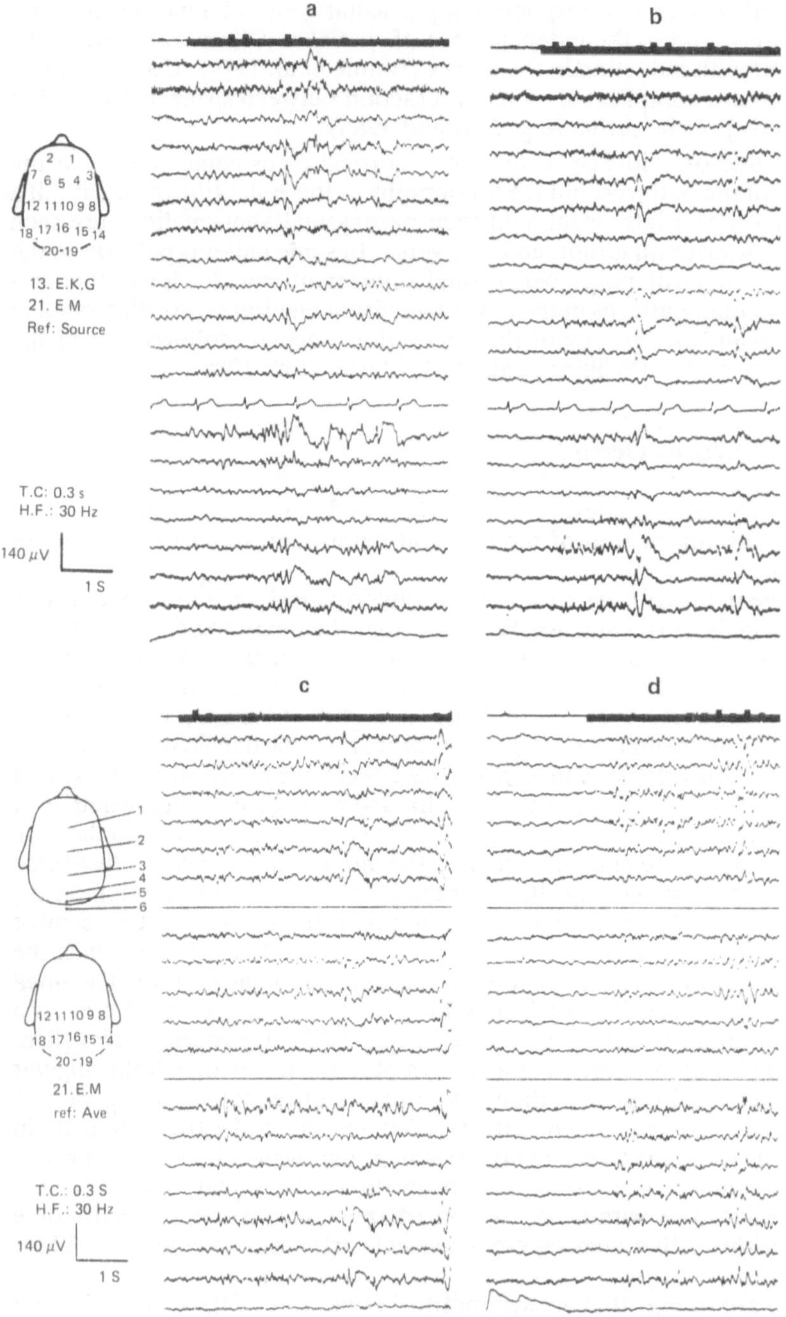

Fig. 10.2a–d. Examples of the paroxysmal EEG activity recorded shortly before and during the presentation of a pattern in the left (**a**), the right (**b**), the upper (**c**) and the lower (**d**) visual half field. Recordings are from one patient. A weighted average ("source") was used as reference in recordings **a** and **b**, whereas in recordings **c** and **d**, which used unconventional electrode placements, average reference was used. *T.C.*, time constant; *H.F.*, high frequency filter.

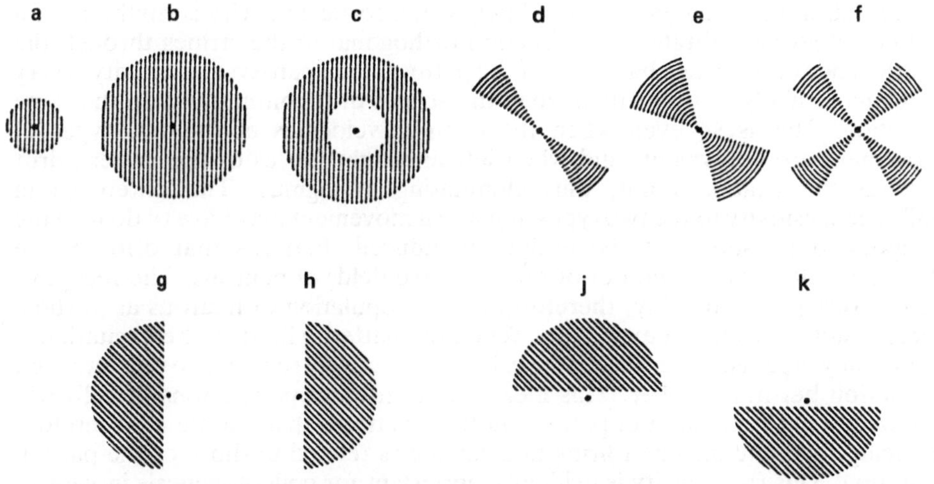

Fig. 10.3a–k. The physiological excitation from patterns stimulation depends upon pattern size. The cortical projection of patterns **a** and **c** is similar and less than **b**, although the area of **c** is larger than **a**. Patterns **e** and **f** have similar area and cortical projection. Patterns **g** and **h** project to the right and left hemispheres, respectively, and patterns **j** and **k** to both hemispheres.

fields (such as patterns j or k in Fig. 10.3). This result may be explained if both hemispheres act independently in the induction of seizure activity: patterns j and k stimulate both hemispheres, but the critical mass in each hemisphere may or may not be exceeded. The independence of the hemispheres is emphasised by the fact that if a bilateral pattern is made by combining patterns g and h in Fig. 10.3 together, then the probability of paroxysmal abnormalities is not increased more than might be expected from the combination of chance factors (Wilkins et al. 1981).

A third line of evidence for the independence of the hemispheres comes from the topography of the EEG response. The topography of the EEG activity changes with the region of visual field stimulated. When small patterns are used (i.e. those that are just sufficiently large to induce a focal paroxysmal response) patterns in the left visual half field produce activity over the right posterior quadrant, and patterns in the right visual field produce activity over the left. When the pattern size is increased, the activity can be more widespread and involve anterior regions (Wilkins et al. 1981; Darby et al. 1986).

In some patients paroxysmal abnormalities are far more likely to appear from stimulation of one lateral visual field than the other, and in these patients the response to diffuse intermittent light shows a slight asymmetry of amplitude consistent with a greater hyperexcitability of one cerebral hemisphere (Binnie et al. 1981).

From the above evidence it is possible to conclude that seizures occur when normal physiological excitation within the visual cortex of one hemisphere exceeds a critical mass. The induction must occur at some point in the visual system before there are appreciable interhemispheric interactions, presumably in the posterior visual areas.

The third factor influencing the occurrence of seizures has to do with synchronisation. Seizures are more likely when cortical activity is rhythmic. If a pattern of stripes vibrates in a direction orthogonal to the stripes through the width of one cycle or one half cycle of the pattern, then paroxysmal activity is very much more likely to occur than when the stripes drift continuously in the same direction. This is so even when the contour velocities of the two types of movement are equivalent, and when left and right halves of the pattern drift towards the fixation point, thus eliminating nystagmus. The difference in epileptic sensitivity to the two types of pattern movement may have to do with the temporal organisation of the excitation induced. Patterns that drift in one direction will move in and out of the receptive fields of neurons. The receptive fields overlap. Presumably, therefore, in the population of neurons as a whole there results a sustained excitation. When the pattern vibrates, the excitation is temporally organised, because some cells are selectively sensitive to one direction of motion but not another. This means that different populations of cells will respond to each direction of pattern motion. Patterns that vibrate are therefore inducing rhythmic bursts of firing at frequencies related to those of the pattern vibration. This rhythmicity is evidently important for epileptogenesis in view of the very much greater sensitivity to vibrating than to drifting patterns. Stationary patterns are less epileptogenic than those that vibrate, but more epileptogenic than those that drift, suggesting that the eye movements that occur during fixation may play a role in synchronising cortical activity (Binnie et al. 1981).

Why Do Seizures Occur?

What is the underlying pathogenesis and why does physiological excitation have its disastrous effects? The answer is quite uncertain but may have to do with a diffuse imbalance of excitatory and inhibitory processes. The animal literature, particularly that concerning the photosensitive baboon, demonstrates pharmacological evidence converging towards the view that inhibitory (GABA-ergic) processes are impaired, but only minimally so. This evidence is now offset by arguments for a role of excitatory processes (see Chap. 11, p. 80). Whatever the origin of the neurochemical imbalance, the imbalance is not sufficient to interfere with normal function (remember that vision is normal inter-ictally), provided that sensory stimulation does not give rise to a massive excitation. If the level of excitation within the nerve network becomes excessive, however, the discharge may evolve because the inhibitory processes are insufficient to prevent it. Inhibitory inter-neurons are shared, and it may be only when the density of excitation within the network as a whole becomes excessive that this sharing of inhibition becomes critical (Meldrum and Wilkins 1984).

Visual Discomfort

Conventional wisdom has it that only 4% of patients with epilepsy are liable to visually induced seizures, and only about 5% of patients show a photoconvulsive response in the EEG when exposed to intermittent light. However, the photoconvulsive response is an extreme reaction, and it will now be argued that less extreme reactions to provocative visual stimulation are not only possible, but indeed commonplace. Sensitivity to light may be responsible for many complaints of "eye-strain" and headaches.

The argument begins with the observation that people find some patterns unpleasant to look at. Figure 10.4 is one such example. Some people report seeing illusions in patterns such as this: illusions of colour, shape and motion. The illusions can be measured by giving individuals a check list and having them look at the pattern for a few seconds and then tick off the illusions they saw. Generally speaking, the more illusions people see, the more unpleasant they find the pattern. Some individuals are far more susceptible than others to these illusions,

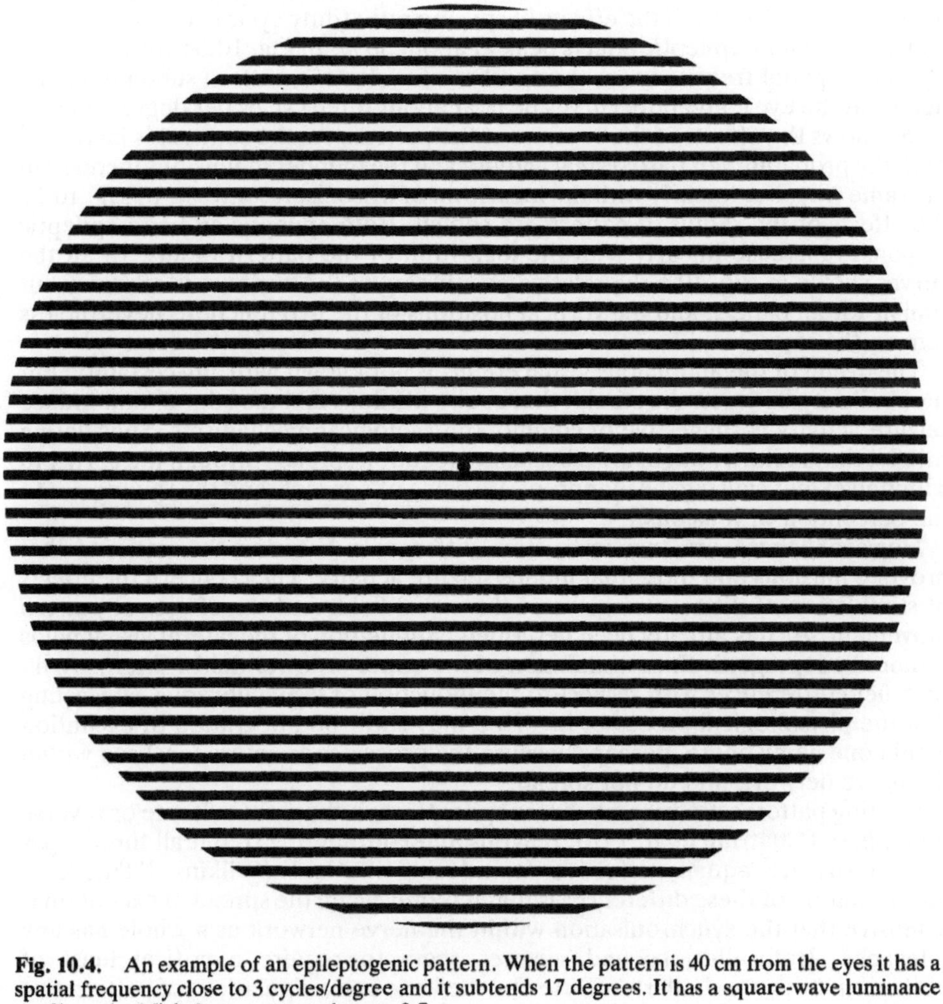

Fig. 10.4. An example of an epileptogenic pattern. When the pattern is 40 cm from the eyes it has a spatial frequency close to 3 cycles/degree and it subtends 17 degrees. It has a square-wave luminance profile and a Michelson contrast close to 0.7.

and there is a relationship between the illusions people see and the headaches they suffer. People who see many illusions tend to suffer frequent headaches. The correlation between the number of illusions people report and the number of headaches they have acounts for about 20% of the variance. The relationship obtains only for certain spatial frequencies: those that are epileptogenic (Wilkins and Nimmo-Smith 1984). If people observe a pattern with the appropriate spatial frequency daily, some report more illusions on days when they have a headache (Nulty et al. 1987). If they get a headache on one side of the head consistently, the illusions tend to predominate in one lateral visual field (Wilkins and Nimmo-Smith 1984).

The illusions are evoked by patterns very similar to those that induce seizures. In Fig. 10.5a are shown the effects of the length of line contour. In a chequerboard pattern, where the length of the check has been progressively stretched, the probability of paroxysmal activity, and the number of illusions of colour, shape and motion both increase similarly with the length/width ratio of the checks. In Fig. 10.5b the effects of spacing ratio (duty cycle) are shown. Both epileptic and non-epileptic effects peak at about 50%. In Fig. 10.5c are shown the effects of spatial frequency (or the number of cycles of a pattern subtending one degree at the eye), and both curves peak at about three cycles per degree. Figure 10.5d shows the effects of the contrast of the pattern. As the contrast is increased both the probability of paroxysmal activity and the number of illusions increase in the same way, i.e. linearly with the logarithm of Michelson contrast. In Fig. 10.5e the effects of the radius of pattern are shown. Both epileptic and non-epileptic responses increase linearly with the logarithm of the pattern radius. In all the curves shown in Fig. 10.5 the relative positioning of the two axes is arbitrary but similar for all panels, and the vertical positions of the two functions nevertheless remain the same.

The origin of the anomalous visual effects is not known with any certainty, but the similarity between the stimuli that evoke illusions and the stimuli that induce seizure activity is suggestive of common physiological mechanisms. Speculating briefly, the similarity might arise because excitation spreads through the cortex to a limited extent causing the inappropriate firing of neurons, and thus giving rise to the perception of illusions.

However, there are two instructive differences between the patterns that provoke illusions and those that induce seizure activity. They concern the effects of stimulation of different regions of the visual field, and the effects of pattern movement. As has already been described, stimulation of the lateral visual fields is more epileptogenic than stimulation of the upper or lower fields. There are no such field differences with respect to the induction of visual illusions, suggesting that the physiological processes involved may show no integration of excitation within one hemisphere, perhaps because the processes are more localised within the nerve network and do not spread.

Drifting patterns are very much less epileptic than those that vibrate or reverse their phase (Chatrian et al. 1970; Jeavons and Harding 1975), but all three types of patterns are equally effective at inducing illusions (Wilkins 1986). One interpretation of these differences is that it is only when the spread of excitation is extensive that the synchronisation within the nerve network as a whole has any relevance. Both illusions and seizures seem to require a critical mass of excitation: both are similarly affected by the various manipulations of pattern size shown in Fig. 10.3.

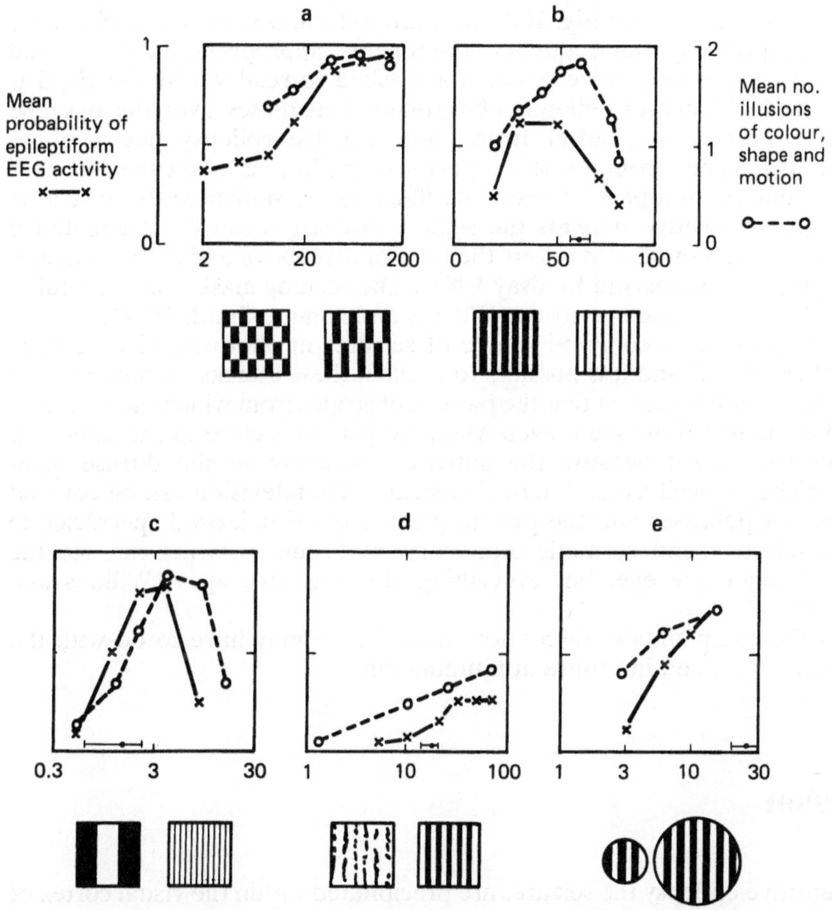

Fig. 10.5a–e. The influence of various pattern characteristics on the illusions seen by normal observers, and the occurrence of paroxysmal epileptiform activity in the EEG of patients with photosensitive epilepsy. Each pattern parameter was manipulated independently, the values of the other parameters remaining close to those at which illusions and paroxysmal activity were maximally likely. Schematic diagrams of the stimuli are shown beneath the x-axis. The mean number of illusions reported by a group of observers is shown by the *broken curves*. The *solid curves* show the mean proportion of randomised presentations associated with the occurrence of epileptiform EEG activity for a group of photosensitive patients (**b–e**) or one such patient (**a**). These dependent variables are shown as functions of **a** the height/width ratio of line segments in patterns of elongated checks; **b** the percentage of spatial cycle occupied by bright bars; **c** the spatial frequency (number of cycles in one degree subtended at the eye); **d** Michelson contrast, the ratio (L2-L1)/(L2+L1), expressed as a percentage, where L2 and L1 are the luminance of the bright and dark bars, respectively; **e** pattern radius (in degrees subtended at the eye). The *horizontal points* and *bars* near the abscissae show the mean and SD of values typical of printed text when the latter is considered as a pattern of horizontal stripes. Note that the relative scaling of the two y-axes is arbitrary, but constant for all graphs. (Based on Wilkins et al. 1980; Wilkins and Nimmo-Smith 1984)

Practical Aspects

The successive lines of printed text comprise a pattern of stripes, and for certain texts the characteristics of this pattern fall within the range of parameters that lead to seizures. As shown in Fig. 10.5, the pattern from text has the appropriate size, the appropriate spatial frequency components, appropriate duty cycle and contrast. If very photosensitive patients are asked to read whilst the EEG is recorded, the incidence of epileptic abnormalities increases over the baseline rate. People who do not suffer from photosensitive epilepsy but who are susceptible to illusions from stripes, report seeing illusions when they fixate a letter in the middle of a page of text. The illusions are similar to those seen in stripes. In photosensitive patients the seizure discharges can be attenuated if patients read using a mask that covers the lines of text above and below the ones they are reading (Wilkins and Lindsay 1985). The reading mask is also useful in preventing headaches and eye strain (Wilkins and Nimmo-Smith 1984).

Television provides a common source of seizures in photosensitive patients (Wilkins et al. 1979b) and it is possible to prevent these attacks by reducing the size of the television screen so that the pattern of stripes from which the picture is composed is too fine to be seen, even when the patient is close to the screen. If this proves insufficient because the patient is sensitive to the diffuse high-frequency flicker as well as the interlacing stripes, the television can be covered with a sheet of polariser and the patient given crossed polarised spectacles to produce a selective and cosmetic monocular occlusion: patients can see the television through one eye, but everything else with two eyes (Wilkins and Lindsay 1985).

Some of the complaints of discomfort from VDUs may have to do with the combination of text and intermittent illumination.

Conclusion

In photosensitive epilepsy the seizures are precipitated within the visual cortex of one cerebral hemisphere and can remain sustained within it. They occur when normal physiological excitation within that hemisphere exceeds a critical mass, particularly when the excitation is synchronised. Less extreme mechanisms may be responsible for visual discomfort. Both epileptic and non-epileptic types of visual sensitivity depend on a critical mass of excitation.

References

Binnie CD, Darby CE, Wilkins AJ (1979) Pattern-sensitivity: the role of movement. In: Lechner H, Aranibar A (eds) Proceedings of the second European congress of electroencephalography and clinical neurophysiology. Elsevier, Amsterdam, pp 650–655

Binnie CD, Wilkins AJ, de Korte RA (1981) Interhemispheric differences in photosensitivity: II. Intermittent photic stimulation. J Electroencephalogr Clin Neurophysiol 52:469–472

Chatrian GE, Lettich E, Miller LH, Green JR (1970) Pattern-sensitivity epilepsy Part 1. An electrographic study of its mechanisms. Epilepsia 11:125–149

Darby C, Park D, Smith A, Wilkins A (1986) EEG characteristics of epileptic pattern sensitivity and its relation to the nature of pattern stimulation and the effects of sodium valproate. Electroencephalogr Clin Neurophysiol 63:517–525

Hubel DH, Wiesel TN (1979) Brain mechanisms of vision. Sci Am 241:130–144

Jeavons PM, Harding GFA (1975) Photosensitive epilepsy. Heinemann, London

Masland RH (1986) The functional architecture of the retina. Sci Am 255 (6):90–99

Meldrum BS, Wilkins AJ (1984) Photosensitive epilepsy: integration of pharmacological and psychophysical evidence. In: Schwartzkroin P, Wheal HW (eds) Electrophysiology of epilepsy. Academic, London, pp 51–77

Nulty DD, Wilkins AJ, Williams JM (1987) Mood, pattern sensitivity and headache: a longitudinal study. Psychol Med 17:705–713

Wilkins AJ (1986) What is visual discomfort? Trends Neurosci 9 (8):343–346

Wilkins AJ, Lindsay J (1985) Common forms of epilepsy: physiological mechanisms and techniques for treatment. In: Pedley TA, Meldrum BS (eds) Recent advances in epilepsy II. Churchill Livingstone, Edinburgh, pp 239–271

Wilkins AJ, Nimmo-Smith I (1984) On the reduction of eye-strain when reading. Ophthalmic Physiol Opt 4:53–59

Wilkins AJ, Darby CE, Binnie CD (1979a) Neurophysiological aspects of pattern-sensitive epilepsy. Brain 102:1–25

Wilkins AJ, Darby CE, Stefansson SF, Jeavons PM, Harding GFA (1979b) Television epilepsy: the role of pattern. J Electroencephalog Clin Neurophysiol 47:163–171

Wilkins AJ, Binnie CD, Darby CE (1980) Visually-induced seizures. Prog Neurobiol 15:85–117

Wilkins AJ, Binnie CD, Darby CE (1981) Interhemispheric differences in photosensitive epilepsy I: pattern sensitivity thresholds. J Electroencephalogr Clin Neurophysiol 5:461–468

Chapter 11

The Basal Ganglia and the Development and Motor Expression of Partial Seizures

B. S. Meldrum

Introduction

Hughlings Jackson's views on hierarchies have come into question in the last 100 years, whereas his anatomical observations in epilepsy have stood the test of time. He provided not only the classic clinical descriptions of focal motor seizures and of temporal lobe epilepsy but also clear evidence concerning the anatomical localisation of the abnormal discharges (Jackson and Beevor 1889; Taylor 1931/32). The focal origin of the abnormal discharges giving rise to the "dreamy states" or psychomotor seizures was presumed to be the anterior and medial aspect of the temporal lobe on the basis of two cases showing space-occupying lesions at post-mortem. The interpretation of the site of origin and spread of focal motor seizures relied heavily on the experimental studies of David Ferrier.

This review will consider more recent information concerning pathways that appear to be important in the spread of seizure activity and in the clinical manifestations of these two forms of partial epilepsy. The basal ganglia operate not only as a relay system for the abnormal discharges but also, through their output, appear to modulate seizure threshold, particularly within the limbic system.

Definitive evidence for the role of the basal ganglia in focal and generalised clonic or tonic seizures was provided by Hayashi (1952, 1953). In a remarkably comprehensive series of experiments on monkeys and dogs, he injected excitant agents focally in the cortex or in subcortical nuclei. He tested more than 200 compounds in this way, and showed for the first time that dicarboxylic amino acids such as glutamate and aspartate induce focal seizures. Most commonly he injected nicotine, strychnine, picrotoxin or pentylenetetrazol, either into the motor cortex and adjacent areas or into subcortical structures (thalamus and basal ganglia). Clonic convulsions could be evoked by the injection of nicotine in the globus pallidus or the substantia nigra reticulata. The clonic convulsions induced by intracortical injections of nicotine can be blocked by bilateral lesions of either the globus pallidus or the rostral midbrain (substantia nigra). These results were summarised by Hayashi in the form of a diagram of the conducting path for clonic convulsions (Hayashi 1953) (Fig. 11.1). They led Hayashi to

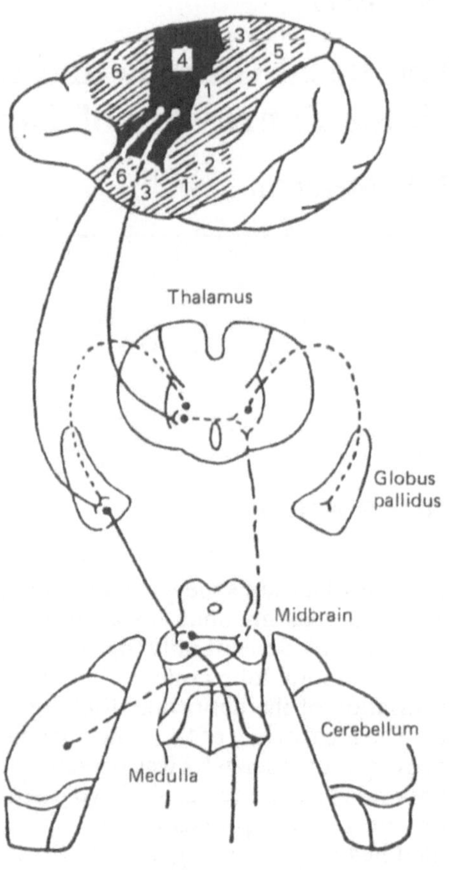

Fig. 11.1. Levels which elicit clonic convulsions and the presumed descending pathway for cortically induced clonus, as defined by Hayashi on the basis of experiments in dogs and monkeys. (Modified from Hayashi 1953)

propose that focal epilepsy could be treated by lesioning the globus pallidus or its output pathway. This procedure ("lenticotomy") prevented not only the focal motor seizures but also the secondary generalisation and loss of consciousness in some patients (Hayashi 1952). A stereotactic procedure for lesioning the globus pallidus was also employed by neurosurgeons in the United Kingdom (Gillingham 1980).

Various Italian authors (Mutani, La Grutta and colleagues) have used depth electrodes in the cat to study the effect of focal stimulation in the basal ganglia on cortical or limbic seizure activity. Prior high-frequency electrical stimulation of the head of the caudate nucleus decreases electrically induced afterdischarges in the amygdala or hippocampus (La Grutta et al. 1971, 1985, 1986). Electrically induced afterdischarges or seizures in the amygdala are more powerfully suppressed by conditioning stimulation of the globus pallidus or substantia nigra than of the striatum (Amato et al. 1982). Lesions in the globus pallidus or substantia nigra decrease the inhibitory effect on amygdala discharges of striatal stimulation, implying that the basal ganglia output pathway is responsible for this inhibitory effect (Amato et al. 1982). By using the intravenous administration of penicillin to induce "inter-ictal" spikes in the hippocampus, somewhat different results have been obtained (Sabatino et al. 1986). Stimulation of the substantia nigra pars compacta suppresses the spikes, whereas stimulation of the globus pallidus pars interna facilitates their evolution into seizure discharges. Destruction of the medial septal nucleus prevents both these effects (Sabatino et al. 1986), suggesting that a pallido-habenula-septal pathway could be providing a means for basal ganglia output to control hippocampal excitability.

Focal Cortical Seizures

Interest in the pattern of anatomical spread of seizure activity was greatly enhanced about 10 years ago following the introduction of methods that allowed the visualisation and quantitative measurement of regional changes in cerebral blood flow and glucose metabolism. Notably Caveness and his colleagues studied focal motor seizures (induced in monkeys by intracortical injection of penicillin) using 14C-antipyrine or 14C-deoxyglucose and autoradiography of frozen sections. They observed an increase in regional blood flow and glucose utilisation that principally involved ipsilaterally the putamen, globus pallidus and substantia nigra, and the ventrolateral and ventromedial thalamic nuclei, and contralaterally the cerebellar hemispheres and dentate nucleus (Ueno et al. 1975; Caveness et al. 1980). With seizure activity confined to the face–hand area of the neocortex somatotopically preserved activation is seen in the putamen, globus pallidus and substantia nigra (Hosokawa et al. 1983).

Kindled Limbic Seizures and the Basal Ganglia

Goddard et al. (1969) described an animal model of limbic seizures ("kindling") dependent on daily or intermittent electrical stimulation of susceptible brain areas (amygdala, hippocampus, entorhinal or piriform cortex) at a strength which initially triggers only an afterdischarge but eventually initiates a seizure discharge that involves the limbic system bilaterally and is associated with a characteristic motor sequence (beginning with gustatory automatisms, progressing to clonus of the forelimbs, and finally rearing and falling). Activation of the substantia nigra during kindled seizures has been shown both by regional metabolic studies employing 14C-deoxyglucose (Engel et al. 1978) and by electrical recording of nigral unit activity (Bonhaus et al. 1986). The focal injection of GABA-ergic agents (γ-vinylGABA, which inhibits GABA-transaminase, or muscimol, a GABA A agonist) bilaterally into the substantia nigra partially or completely suppresses amygdala kindled seizures (LeGal La Salle et al. 1983; McNamara et al. 1984). Bilateral destruction of the substantia nigra (by focal injection of a neurotoxin) also suppresses the limbic seizures (McNamara et al. 1984). Bilateral electrical stimulation of the substantia nigra pars reticulata (at a strength less than that inducing tonic seizures) prior to stimulation of the amygdala delays and diminishes kindled responses (Morimoto and Goddard 1987). If the electrical stimulation of the substantia nigra pars reticulata is primarily activating intrinsic inhibitory terminals, then all these results can be interpreted as showing that decreasing the output from the substantia nigra pars reticulata prevents the build-up of limbic seizure activity.

More detailed evidence on this point has been obtained using a chemically induced model of limbic seizures.

Limbic Seizures Induced in Rats by Pilocarpine

High doses of the muscarinic agonist pilocarpine given systemically to rats induce sustained seizures whose motor pattern is similar to that of kindled limbic seizures induced by amygdala stimulation (described above). The motor and electrical seizure activity is commonly sustained for 3–5 h and leads to neuronal loss in the hippocampus, amygdala, thalamus, neocortex and substantia nigra (Turski et al. 1983). This seizure syndrome may not be an optimal model for pharmacological studies of complex partial seizures, as diphenylhydantoin and carbamazepine are relatively ineffective at suppressing the seizures (and the brain damage), whereas clonazepam, phenobarbital and valproate are effective (Turski W. A. et al. 1987). However, these seizures are very useful for establishing the pathways involved in the initiation and development of limbic seizures.

We have studied the effect of focal injections of muscimol, (a GABA agonist), of *N*-methyl-D-aspartate (a selective agonist at one subtype of glutamate receptor), and of two selective excitatory amino acid antagonists (2-amino-7-phosphonoheptanoic acid, 2-APH, acting on the NMDA receptor, and γ-D-glutamylamino-methylsulphonate, GAMS, acting with partial selectivity on the kainate receptor) (Meldrum et al. 1988) (Fig. 11.2).

Focal injection of muscimol or 2-APH into the substantia nigra pars reticulata completely prevents the motor limbic seizures and the secondary pathology induced by a convulsant dose of pilocarpine (Turski et al. 1986). Focal EEG recordings show that isolated spikes occur in the hippocampus following the systemic injection of pilocarpine, but they fail to develop into a sustained discharge. Injection of NMDA into the substantia nigra pars reticulata can convert a subconvulsant dose of pilocarpine into a fully convulsant dose. These experiments were interpreted as showing that decreasing the output from the substantia nigra pars reticulata has an inhibitory effect on the limbic system, whereas enhancing its output facilitates seizure activity in the limbic system. The outputs involved are probably GABA-ergic, relaying in either the superior colliculus, the ventral tegmental area or the thalamus (ventromedial nucleus).

The other major output nucleus for the basal ganglia is the globus pallidus. In the rat the entopeduncular nucleus is homologous with the internal part of the globus pallidus in the primate or cat. Focal injection of muscimol or of 2-APH potently protects against pilocarpine-induced seizures. Indeed, the effective dose of 2-APH at this site is some 100-fold lower than that required in the substantia nigra (Patel et al. 1986). Interestingly, γ-D-GAMS, which acts preferentially on the kainate receptor, is nearly as effective as 2-APH in the substantia nigra pars reticulata but very much less effective in the entopeduncular nucleus (De Sarro et al. 1986). An important output of the entopeduncular nucleus is to the lateral habenula, which in turn relays to the thalamus (including the mediodorsal nucleus). Protection against limbic seizures induced by pilocarpine is seen after

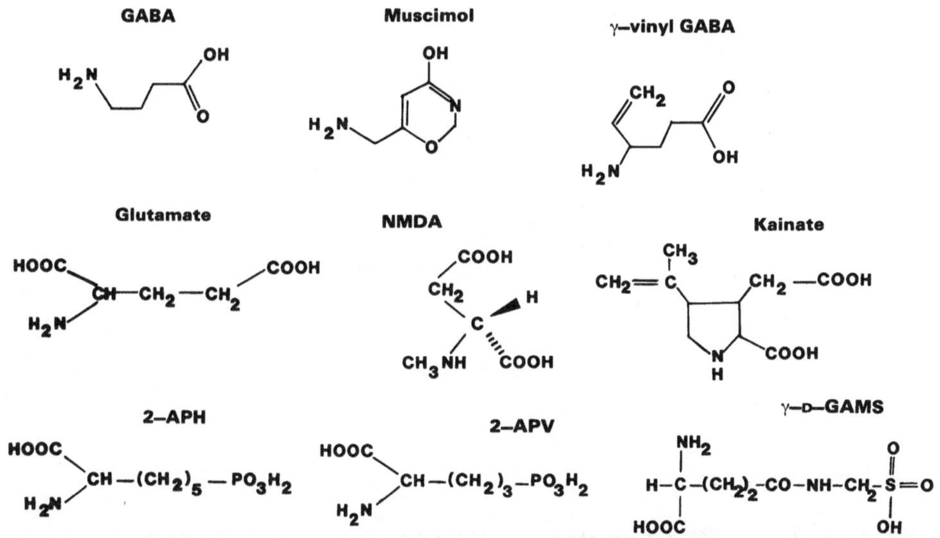

Fig. 11.2. Molecular formulae of agents acting on inhibitory or excitatory neurotransmission, and used to study the role of deep brain nuclei in seizure phenomena. Muscimol is an agonist acting on GABA A receptors; γ-vinyl-GABA (vigabatrin) is an irreversible inhibitor of GABA-transaminase. *N*-methyl-D-aspartate (*NMDA*) is an agonist acting on a subtype of glutamate or excitatory amino acid receptor. 2-amino-7-phosphonoheptanoic acid (*2-APH*) and 2-amino-5-phosphorovaleric acid (*2-APV*) are potent specific antagonists acting on the NMDA receptor. γ-D-glutamylamino-methylsulphonate γ-D-GAMS is an antagonist with a preferential action on the "kainate receptor" (another sub-type of excitatory amino acid receptor).

focal injection of 2-APH into either the lateral habenula or the mediodorsal nucleus of the thalamus (Fig. 11.3).

However, the site which is most sensitive to the protective action of 2-APH lies deep in the prepiriform cortex (Fig. 11.3) (Millan et al. 1986). This area has been described by Piredda and Gale (1985) as a crucial epileptogenic site on the basis of its high sensitivity to the epileptogenic action of various convulsants including bicuculline (a GABA antagonist), kainic acid, or carbachol (cholinergic agonist). It is possible that pilocarpine limbic seizures are initiated here and spread to the amygdala and hippocampus.

N-methyl-D-aspartate is convulsant when given systemically and is proconvulsant in the pilocarpine-induced limbic seizure model when injected into the substantia nigra pars reticulata. When injected focally into the ventral caudate/putamen, however, it is anticonvulsant (Turski L. et al. 1987). This protective effect could be abolished by the simultaneous microinjection of GABA antagonists into the substantia nigra or entopeduncular nucleus. Thus it is probably that NMDA protects against limbic seizures by activating the GABA-ergic inhibitory pathways that run from the striatum to the nigra and entopeduncular nucleus.

These experiments provide clear evidence that the output of the basal ganglia tonically influences excitability within the limbic system. This tonic activity apparently facilitates the development of seizure activity; reducing it suppresses seizure development. The pathways involved are clearly polysynaptic and involve relays in the habenula and thalamus. And possibly also the medial septal nucleus and perhaps the tegmentum.

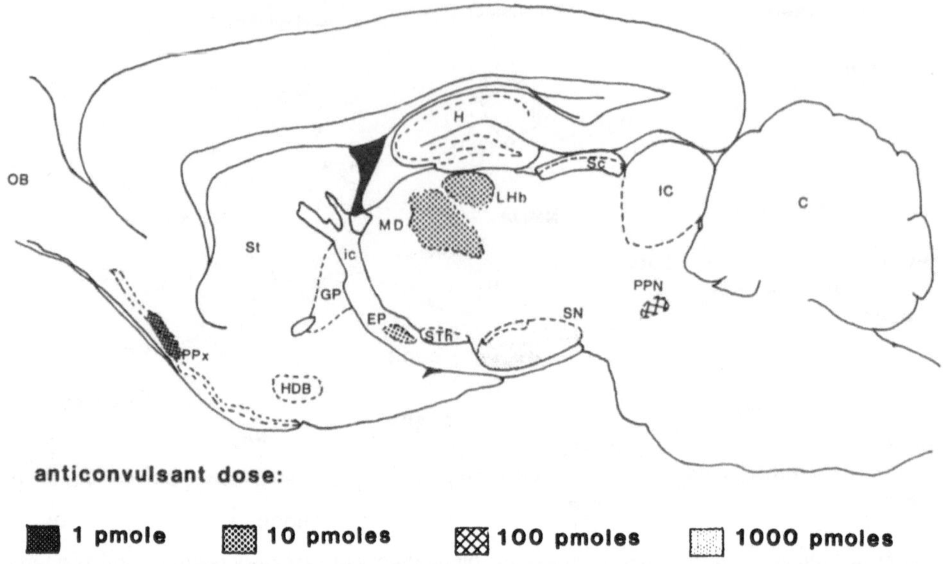

Fig. 11.3. Coronal diagram of the rat brain showing sites at which focal injections of 2-APH have an anticonvulsant action against limbic seizures induced by systemic injection of pilocarpine. The threshold dose producing an anticonvulsant effect is indicated by the pattern of cross-hatching. The prepiriform cortex (*PPx*) is the most sensitive region; the lateral habenula (*LHb*), the mediodorsal nucleus of the thalamus (*MD*) and the entopeduncular nucleus (*EP*) are the next most sensitive. The pedunculopontine nucleus (*PPN*) is less sensitive and the substantia nigra pars reticulata (*SN*) is the least sensitive. (Reproduced from Patel et al. 1987, with permission).

Myoclonus of Striatal Origin

In the rat, focal contralateral myoclonus follows the injection into the striatum of GABA antagonists (either acting directly, such as picrotoxin and bicuculline, or indirectly by blocking GABA synthesis, such as allylglycine (Tarsy et al. 1978). This effect is facilitated by cortical lesions, and is prevented by lesions of the globus pallidus. Pallidal lesions do not prevent myoclonus induced by intracortical injection of picrotoxin (Patel and Slater 1987). A circuit involving striatal outputs to globus pallidus, entopeduncular nucleus and substantia nigra with a relay to the ventral thalamus and then to motor cortex appears to be involved in striatally evoked myoclonus.

Basal Ganglia in Neonatal Seizures

The pattern of metabolic activation during focal motor seizures in newborn monkeys is similar to that occurring in pubescent monkeys, but more ipsilateral deep structures show an increased metabolism (Kato et al. 1980). These include the ventrocaudal parts of the putamen and the globus pallidus, the substantia nigra and the ventroposterolateral nucleus of the thalamus. Experiments in rat pups suggest that the functional relationships are different, however, in the newborn period. Kindled seizures in rat pups are not associated with enhanced glucose metabolism in the globus pallidus and substantia nigra (Ackermann et al. 1982). Furthermore, bilateral injection of muscimol into the substantia nigra decreases the motor component of flurothyl seizures in adult rats but facilitates such seizures in 15-day-old pups (Moshé and Albala 1984).

Conclusions

1. *The basal ganglia are a relay pathway for seizure activity.* Certain motor manifestations of seizure activity depend on information relayed through the globus pallidus and the substantia nigra, as first shown by Hayashi (although we now see these nuclei as parallel outputs, rather than as consecutive relays). Relays in the pedunculopontine nuclei or the midbrain reticular formation provide access to the final common path.

2. *The basal ganglia provide tonic modulation of limbic seizure threshold.* This conclusion derives from the experiments with focal injections of excitatory and inhibitory agents in the kindling model and in pilocarpine-induced limbic seizures. The polysynaptic pathways involved may include relays in the habenula, thalamus and medial septum. This effect may be related to a functional role of the basal ganglia in integrating facial motor activity and sensory inputs.

References

Ackermann RF, Moshé SL, Albala BJ, Engel J (1982) Anatomical substrates of amygdala kindling in immature rats demonstrated by 2-deoxyglucose autoradiography. Epilepsia 23:434–435

Amato G, Crescimanno G, Sorbera F, La Grutta V (1982) Relationship between the striatal system and amygdaloid paroxysmal activity. Exp Neurol 77:492–504

Bonhaus DW, Walters JR, McNamara JO (1986) Activation of substantia nigra neurons: role in the propagation of seizures in kindled rats. J Neurosci 6:3024–3030

Caveness WF, Kato M, Malamut BL, Hosokawa S, Wakisaka S, O'Neill R (1980) Propagation of focal motor seizures in the pubescent monkey. Ann Neurol 7:213–221

De Sarro G, Meldrum BS, Reavill C (1985) Anticonvulsant action of 2-amino-7-phosphonoheptanoic acid in the substantia nigra. Eur J Pharmacol 106:175–179

De Sarro G, Patel S, Meldrum BS (1986) Anticonvulsant action of a kainate antagonist γ-D-glutamyl aminomethylsulphonic acid injected focally into the substantia nigra and entopeduncular nucleus. Eur J Pharmacol 132:229–236

Engel J, Wolfson L, Brown L (1978) Anatomical correlates of electrical and behavioral events related to amygdaloid kindling. Ann Neurol 3:538–544

Gillingham FJ, Watson WS, Chung S, Yates L (1980) Central brain lesions for the control of intractable epilepsy. In: Watson JA, Penry JK (eds) Advances in epileptology: the Xth epilepsy international symposium. Raven, New York, pp. 251–255

Goddard GV, McIntyre DC, Leech CK (1969) A permanent change in brain function resulting from daily electrical stimulation. Exp Neurol 25:295–330

Hayashi T (1952) A physiological study of epileptic seizures following cortical stimulation in animals and its application to human clinics. Jpn J Pharmacol 3:46–64

Hayashi T (1953) The efferent pathway of epileptic seizures for the face following cortical stimulation differs from that for limbs. Jpn J Pharmacol 4:306–321
Hosokawa S, Kato M, Kuroiwa Y (1983) Topographical distribution of propagation of seizure activity in the basal ganglia during focal motor seizures in the monkey. Neurosci Lett 38:29–33
Jackson JH, Beevor CE (1889) Case of tumour of the right temporo-sphenoidal lobe bearing on the localization of the sense of smell and on the interpretation of a particular variety of epilepsy. Brain 12:346–357
Kato M, Malamut BL, Caveness WF, Hosokawa S, Wakisaka S, O'Neill RR (1980) Low cerebral glucose utilization in newborn and pubescent monkeys during focal motor seizures. Ann Neurol 7:204–212
La Grutta V, Amato G, Zagami MT (1971) The importance of the caudate nucleus in the control of convulsive activity in the amygdaloid complex and the temporal cortex of the cat. Electroencephalogr Clin Neurophysiol 31:57–69
La Grutta V, Sabatino M, Gravante G, La Grutta G (1985) Effects of caudate nucleus on paroxysmal activity in hippocampus of cat. Electorencephalogr Clin Neurophysiol 61:416–421
La Grutta V, Sabatino M, Ferraro G, Liberti G, La Grutta G (1986) Hippocampal seizures and striatal regulation: a possible functional pathway. Neurosci Lett 72:277–282
Le Gal La Salle G, Kajima M, Feldblum S (1983) Abortive amygdaloid kindled seizures following microinjection of γ-vinylGABA in the vicinity of substantia nigra in rats. Neurosci Lett 36:69–74
McNamara JO, Galloway MT, Rigsbee LC, Shin C (1984) Evidence implicating substantia nigra in regulation of kindled seizure threshold. J Neurosci 4:2410–2417
Meldrum B, Millan M, Patel S, De Sarro G (1988) Anti-epileptic effects of focal micro-injection of excitatory amino acid antagonists. J Neurol Transm 72:191–200
Millan MH, Patel S, Mello LM, Meldrum BS (1986) Focal injection of 2-amino-7-phosphonoheptanoic acid into prepiriform cortex protects against pilocarpine-induced limbic seizures in rats. Neurosci Lett 70:69–74
Morimoto K, Goddard GV (1987) The substantia nigra is an important site for the containment of seizure generalization in the kindling model of epilepsy. Epilepsia 28:1–10
Moshé SL, Albala BJ (1984) Nigral muscimol infusions facilitate the development of seizures in immature rats. Dev Brain Res 13:305–308
Patel S, Slater P (1987) Analysis of the brain regions involved in myoclonus produced by intracerebral picrotoxin. Neuroscience 20:687–693
Patel S, Millan MH, Mello LM, Meldrum BS (1986) 2-amino-7-phosphonoheptanoic acid (2-APH) infusion into entopeduncular nucleus protects against limbic seizures in rats. Neurosci Lett 64:226–230
Patel S, Chapman AG, Millan MH, Meldrum BS (1987) Epilepsy and excitatory amino acid antagonists. In: D Lodge (ed) Excitatory amino acids in health and disease. Wiley, London, pp 353–378
Piredda S, Gale K (1985) A crucial epileptogenic site in the deep prepiriform cortex. Nature 317:623–625
Sabatino M, Savatteri V, Liberti G, Vella N, La Grutta V (1986) Effects of substantia nigra and pallidum stimulation on hippocampal interictal activity in the cat. Neurosci Lett 64:293–298
Tarsy D, Pycock CJ, Meldrum BS, Marsden CD (1978) Focal contralateral myoclonus produced by inhibition of GABA action in the caudate nucleus of rats. Brain 101:143–162
Taylor J (ed) (1931/32) Selected writings of John Hughlings Jackson, vols 1 and 2. Hodder and Stoughton, London. Reprinted (1958) Basic Books, New York
Turski L, Cavalheiro EA, Turski WA, Meldrum BS (1986) Excitatory neurotransmission within substantia nigra pars reticulata regulates threshold for seizures by pilocarpine in rats: effects of intranigral 2-amino-7-phosphonoheptanoate and *N*-methyl-D-aspartate. Neuroscience 18:61–77
Turski L, Meldrum BS, Cavalheiro EA, Calderazzo-Filho ZA, Ikonomidou-Turski C, Turski WA (1987) Paradoxical anticonvulsant activity of the excitatory amino acid *N*-methyl-D-aspartate in the rat caudate-putamen. Proc Natl Acad Sci USA 84:1689–1693
Turski WA, Cavalheiro EA, Schwarz M, Czuczwar SJ, Kleinrok Z, Turski L (1983) Limbic seizures produced by pilocarpine in rats: a behavioural, electroencephalographic and neuropathological study. Behav Brain Res 9:315–335
Turski WA, Cavalheiro EA, Coimbra C, Berzaghi MdP, Ikonimidou-Turski C, Turski L (1987) Only certain antiepileptic drugs prevent seizures induced by pilocarpine. Brain Res 434:281–305
Ueno H, Yamashita Y, Caveness WF (1975) Regional cerebral blood flow pattern in focal epileptiform seizures in the monkey. Exp Neurol 47:81–96

Section IV

SENSORY SYSTEMS

Chapter 12

Hierarchies and the Visual System

C. Kennard

Since Jackson proposed his concepts of increasing levels of complexity and hierarchy in the nervous system it has been the visual system which, of all the sensory systems, has appeared to provide the most support. Indeed, at the time Jackson was writing a number of individuals, including Hering and Helmholtz, were suggesting that just as one can break down a visual image into its elements by the process of introspection so the nervous system can build up an image of the external world by analysing these components in a piecemeal way, e.g. the lines that make up a shape, and the wavelengths that form the colours. These analysed components are then brought together at successive levels of the visual pathways to construct a final image of the external world. The implication of this concept was that visual information was serially processed at levels of ever-increasing complexity. Support for this concept was forthcoming from the early neurophysiological studies of the visual pathways performed by Hubel and Wiesel (1962, 1965). It had already been shown that the receptive field of retinal ganglion cells, and neurons in the next staging post, the lateral geniculate nucleus (LGN), are of a centre-surround type. Indeed, the retina itself appears to represent a hierarchical neuronal system with photoreceptors at the lowest level, bipolar cells at an intermediate level and the retinal ganglion cells at the most advanced level. In the striate (Brodmann 1905; area 17) and prestriate cortex (areas 18 and 19) Hubel and Wiesel found neurons with receptive fields of increasing levels of complexity. Simple cells responded to appropriately orientated line stimuli located in their receptive field, and it was clear that such a response could be generated if each cell received inputs from a specific set of ganglion cells. Groups of simple cells analysing a particular orientation appeared to be brought together to produce a complex cell which will respond to angles, and hypercomplex cells showed even more complicted visual field properties. It appeared that by a process of serial processing a hierarchical system of neurons with receptive fields of an ever-increasing complexity were to be found in the visual cortex.

Early anatomical studies also supported a hierarchy in cortical visual processing. Kuypers et al. (1965) found that the primary visual cortex (area 17) projects to a large cortical area which includes areas 18 and 19. This area in turn projects to the inferotemporal cortex, presumed to be an even higher level.

Subsequent laboratory study of the visual system, especially in non-human primates, using a wide range of anatomical, physiological and behavioural techniques, has made it increasingly clear that visual processing is not solely based on a serial, hierarchical system but involves considerable parallel processing and functional specialisation.

The first important observation which contradicted a strictly hierarchical view, was when a number of different prestriate cortical areas were found to receive input from V1 (area 17) (Cragg 1969; Zeki 1969). There are major projections from V1 to three prestriate areas (V2, V3 and V5, also named middle temporal area, MT), as well as several minor projections (review in Maunsell and Newsome 1987). This suggests that a single area of V1, which processes visual inputs from a particular retinal region and hence field of view, must have independent and, therefore, parallel outputs to different prestriate areas. Additional observations which suggest that visual processing is not based on a hierarchical system alone include (1) a small but significant proportion of axons from the LGN projecting to extrastriate areas 18 and 19; (2) backward directed projections from areas 18 and 19 to area 17; and (3) parallel projections originating from different subgroups of retinal ganglion cells, the parvocellular and magnocellular systems, which remain partially segregated until they reach the cortex and other subcortical regions. At the last count the visual cortex had been shown to contain at least 19 separate visual areas with 84 interconnecting pathways (Van Essen 1985). This gives an indication of the considerable complexity of the visual system; yet, despite this, some sort of order and certain rules of connection can be unravelled. A number of different, although interrelated, concepts have been evolved which suggest that it is no longer appropriate to consider the visual pathways as existing in either a hierarchical or a parallel arrangement. Rather it would appear that both these organisational elements are to be found.

Zeki (1978, 1988) has suggested that instead of analysing the visual image the nervous system constructs different aspects of the visual image in different systems, i.e. functional specialisation, and that bringing these submodalities of vision together is not a simple hierarchical process. In particular, the primary visual cortex, area V1, has a major distributive function in addition to its role of analysing the visual field for orientation, and bringing together inputs from the two eyes. It has been proposed that rather than sending information concerning all aspects of vision to each prestriate area, area V1 sends projections, each carrying *different* sets of information, to a number of different areas. In this way each extrastriate visual area is specialised for processing different attributes of the visual scene, e.g. motion, colour and shape.

Neurophysiological recordings from these extrastriate visual areas has certainly revealed some evidence for functional specialisation. Area V3, for example, contains 80% orientation-specific neurons (Felleman and Van Essen 1987), area V4 contains a high proportion of neurons selective for colour (Zeki 1978, 1983) and stimulus form (Desimone and Schein 1987), and area V5 (MT) consists largely of neurons selective for direction, speed and binocular disparity (Zeki 1974; Van Essen et al. 1981; Newsome et al. 1985). In addition to this electrophysiological evidence the effect of localised lesions in specific visual cortical areas lends considerable support to the concept of functional specialisation. Both in non-human primates and man such lesions only give rise to deficits in a particular visual submodality. Lesions in area V4 in monkeys have resulted in either a selective deficit of colour (Wild et al. 1985), or of colour and form (Heywood and Cowey 1985). In man a number of cases have been reported of focal inferior temporal-occipital lesions resulting in defects in colour perception (achromatopsia) (Meadows 1974; Damasio et al. 1980). Area V5 (MT) contains mainly cells selective for moving stimuli, and lesions in this area result in a selective impairment in the ability to track the motion of objects (Newsome et al. 1985; Thurston et al. 1988). A single case report in man describes a patient with bilateral lesions of the superior temporal sulci (presumed to include the homologue of V5 in man) who had normal vision except for impaired motion perception (Zihl et al. 1983).

It must be realised, however, that each functionally selective visual area contains some neurons which lack selectivity for the specific visual modality generally common to the area. Similarly, colour-selective cells are found in areas other than V4 and direction-selective cells in areas other than V5. This suggests that it is unlikely that each visual area plays an exclusive role in analysing any single parameter or visual dimension. Rather the concept of functional streams of processing in the visual cortex has developed. A number of different parallel streams, each processing different types of visual information, arise from area V1 and involve several areas, with each area representing a different level of processing for a particular type of information. For example, Ungerleider and Mishkin (1982) described two streams diverging from the early levels of extrastriate cortex, one leading to visual areas of the parietal lobe and the other to visual areas of the temporal lobe. They further suggested that each stream was concerned with processing a different visual attribute, spatial relationships by the former and visual recognition of objects by the latter. Although these streams are in effect parallel pathways there are undoubtedly interconnections, but more importantly there does appear to be a hierarchy of visual areas within each pathway, which can be deduced by a study of their interconnections (Maunsell and van Essen 1983).

The early experiments of Hubel and Wiesel (summarised, 1977) defining the functional architecture of area V1 suggested a homogeneous distribution of input, albeit topographically organised, with columns of neurons outside layer 4C each processing a specific part of the visual field for a preferred orientation. One preferred orientation column followed another in a remarkably consistent sequence. However, more recent evidence suggests that area V1 is indeed functionally specialised and hence serves as a distributive centre for different visual functions. By using enzyme histochemistry to stain for the metabolic enzyme cytochrome oxidase in layer 2 and 3 it was found that its distribution is not uniform, but occurs in discrete repetitive clusters which, when viewed in tangential sections, take on the appearance of blobs (Horton and Hubel 1981). Combining this technique with electrophysiology reveals a functional segregation in that the cytochrome oxidase blobs contain a concentration of wavelength-selective cells, whereas orientation-selective cells are concentrated in the inter-blob areas (Livingstone and Hubel 1984).

Area V2, which abuts area V1, also contains a representation of the contralateral half of the visual field and has a full set of functional cell types. Here again the information relating to form, colour and motion is kept separate. Staining for cytochrome oxidase has revealed a series of thick and thin stripes in parallel. The thick stripes contain cells which are direction-selective, and the thin stripes contain cells which are wavelength-sensitive.

The forward projections to various other areas from these two visual areas maintain their functional segregations, although this may not be the case for the feedback projections. Analysis of this highly complex network of largely reciprocal connections has led Maunsell and Van Essen (1983) to propose a hierarchical arrangement of visual areas (Fig. 12.1). Based on the laminar distribution of the neurons in each visual area giving rise to these projections and their layer of termination, they have assigned the areas to different levels of processing. Ascending projections, which feed forward from the primary visual area, arise in the superficial cortical layers and terminate primarily in layer 4 and the lower part of layer 3. Descending projections sending back information arise mainly in the deep layers and terminate in the superficial and deep layers. By using these anatomical criteria most connections in the visual cortex can be assigned as forward (ascending) or feedback (descending). The hierarchy was constructed by assigning each area to a level just above that of the highest area which provides an ascending input. This leads to six hierarchical levels for the dozen visual areas involved in the macaque monkey. Projections which have a laminar organisation intermediate between the forward and feedback patterns are considered to be lateral interconnections between areas at the same hierarchical level. It is possible that this laminar organisation is common to all

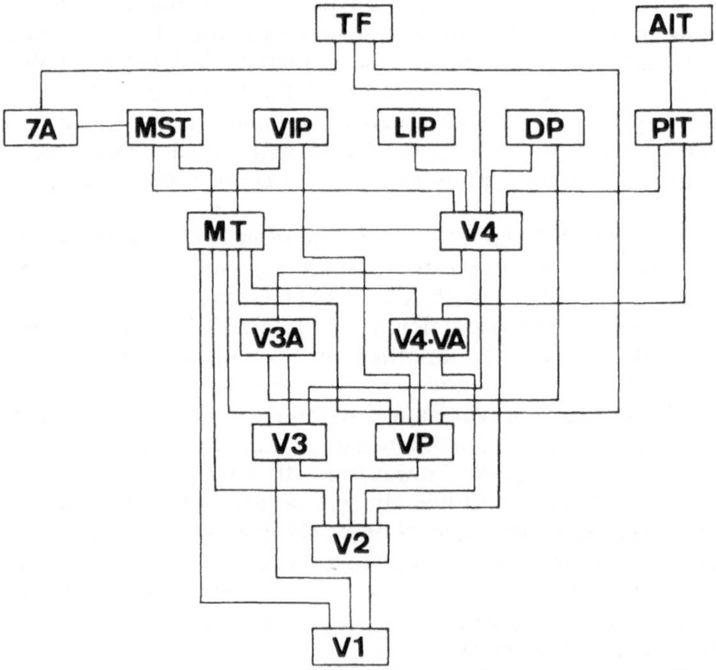

Fig. 12.1. Hierarchy of macaque cortical visual areas. Areas have been assigned to different levels on anatomical criteria by which connections can be assigned as being forward or feedback. Each area is one level above the highest level from which it receives forward input, and below all levels from which it receives feedback. Concomitantly, each area is above all areas to which it sends a feedback projection, and below those to which it sends a forward projection. Abbreviations for visual areas: *AIT*, anterior inferotemporal; *D*, dorsal prelunate; *LIP*, lateral intraparietal; *MST*, medial superior temporal; *MT*, middle temporal; *PIT*, posterior inferotemporal; *TF*, temporofrontal; *VIP*, ventral intraparietal; *VP* ventral posterior. (Reproduced from Maunsell and Newsome 1987, with permission of the Editor of *Annual Review of Neuroscience*)

sensory systems since the multiple areas in the somatosensory system have been shown to be similarly arranged (Friedman 1983).

Recent anatomical studies of these reciprocal connections between visual areas have revealed an interesting way in which visual information is made available to neurons of different functional types (Zeki 1988). The projection from V1 to V5 (MT) arises from clumps of neurons in layers 4B and 6, but the reciprocal descending projection terminates in a continuous manner. This suggests that V5 can modify not only the neurons projecting to V5 but also neurons projecting to other areas, e.g. V3. Similarly the projection from area V2 to area V5 arises from the thick stripes, but the reciprocal projection terminates in the interstripes and thin stripes. Information from the motion area (V5) is therefore not only interacting in a reciprocal manner with movement neurons in the thick stripes of V2, but is also organised so that it can modify the colour-sensitive cells, in the interstripes and thin stripes of V2 which project to V4. Zeki has suggested that this re-entrant system produces reintegration of fractionated visual information which may provide unitary perception of the visual world.

In conclusion, it is apparent that despite the complexity of interconnections between the numerous cortical areas involved in vision there is considerable order. Functional integrity is maintained for different visual modalities and within each functional stream or pathway there lies a hierarchical set of visual areas. However, it is unlikely that "visual perception" for each particular attribute of the visual scene takes place at the highest cortical level of these functional pathways. Rather, the impressive forward and feedback connections in the visual system may ensure a unitary perception of the visual world.

References

Brodman K (1905) Beitrage zur histologischen Lokalisation der Grosshirnrinde. J Psychol Neurol 4:176–226

Cragg BG (1969) The topography of the afferent projections in the circumstrate visual cortex of the monkey studied by the Nauta method. Vision Res 9:733–747

Damasio A, Yamada T, Damasio H, Corbett J, McKee J (1980) Central achromatopsia: behavioral, anatomic and physiologic aspects. Neurology 30:1064–1071

Desimone R, Schein SJ (1987) Visual properties of neurons in area V4 of the macaque: sensitivity to stimulus form. J Neurophysiol 57:835–868

Felleman DJ, Van Essen DC (1987) Receptive field properties of neurons in area V3 of macaque monkey extrastriate cortex. J Neurophysiol 57:889–930

Friedman DP (1983) Laminar patterns of termination of corticocortical afferents in the somatosensory system. Brain Res 273:147–151

Heywood CA, Cowey A (1985) Disturbances of pattern and line discrimination following removal of the colour area in primates. Neurosci Lett 21:S11

Horton JC, Hubel DH (1981) Regular patchy distribution of cytochrome oxidase staining in primary visual cortex of macaque monkey. Nature 292:762–764

Hubel DH, Wiesel TW (1962) Receptive fields, binocular interaction and functional architecture in the cat's visual cortex. J Physiol 160:106–154

Hubel DH, Wiesel TW (1965) Receptive fields and functional architecture in two non-striate visual areas (18 and 19) of the cat. J Neurophysiol 28:229–289

Hubel DH, Wiesel TN (1977) Functional architecture of macaque monkey visual cortex. Proc R Soc Lond [Biol] 198:1–59

Kuypers HGJM, Swarebart MK, Mishkin M, Rosvold HE (1965) Occipitotemporal corticocortical connections in the rhesus monkey. Exp Neurol 11:245–262

Livingstone MS, Hubel DH (1984) Anatomy and physiology of a colour system in the primate visual cortex. J Neurosci 4:309–356
Maunsell JHR, Newsome WT (1987) Visual processing in monkey extrastriate cortex. Annu Rev Neurosci 10:363–402
Maunsell JHR, Van Essen DC (1983) The connections of the middle temporal visual area (MT) and their relationship to a cortical hierarchy in the macaque monkey. J Neurosci 3:2563–2586
Meadows JC (1974) Disturbed perception of colours associated with localised cerebral lesions. Brain 97:615–632
Newsome WT, Wurtz RH, Dursteler MR, Mikami A (1985) Deficits in visual motion processing following ibotenic acid lesions of the middle temporal visual area of the macaque monkey. J Neurosci 5:825–840
Thurston SE, Leigh RJ, Crawford T, Thompson A, Kennard C (1988) Two distinct deficits of visual tracking caused by unilateral lesions of cerebral cortex in man. Ann Neurol 23:266–273
Ungerleider LG, Mishkin M (1982) Two cortical visual systems. In: Ingle DJ, Goodale MA, Mansfield RJW (eds) Analysis of visual behavior. MIT Press, Cambridge, Massachusetts, pp 549–580
Van Essen DC (1985) Functional organisation of the primate visual cortex. In: Peters A, Jones EG (eds) Cerebral cortex 3. Plenum, New York, pp 259–329
Van Essen DC, Maunsell JHR, Bixby JL (1981) The middle temporal area in the macaque: myeloarchitecture, connections, functional properties and topographic organisation. J Comp Neurol 199:293–326
Wild HM, Butler D, Carden D, Kulikowski JJ (1985) Primate cortical area V4 important for colour consistency but not wavelength discrimination. Nature 313:133–135
Zeki S (1969) The secondary visual areas of the monkey. Brain Res 13:197–226
Zeki S (1974) Functional organisation of a visual area in the posterior bank of the superior temporal sulcus of the rhesus monkey. J Physiol (Lond) 236:549–573
Zeki S (1978) Uniformity and diversity of structure and function in rhesus monkey prestriate visual cortex. J Physiol (Lond) 277:273–290
Zeki S (1983) The distribution of wavelength and orientation selective cells in different areas of monkey visual cortex. Proc R Soc Lond [Biol] 217:449–470
Zeki S (1988) Functional specialisation in the visual cortex and its relation to visual agnosias. In: Kennard C, Rose FC (eds) Physiological aspects of clinical neuroophthalmology. Chapman & Hall, London, pp 101–122
Zihl J, von Cramon D, Mai N (1983) Selective disturbance of movement vision after bilateral brain damage. Brain 106:313–340

Chapter 13

Single Fibre Microneurography and Sensation[1]

Å. B. Vallbo

Introduction

The concept of a sensory threshold is a central element in sensory physiology and psychophysics. Yet it is a puzzle that a distinct threshold cannot be demonstrated. A fixed threshold would imply that stimuli above a certain strength are detected with certainty by a human observer while all stimuli below this strength escape detection. Experimental findings, on the other hand, demonstrate that the sensory threshold is a smoothly rising function, i.e. the probability of psychophysical detection gradually rises with stimulus intensity. The plot of the probability of detection as a function of stimulus strength usually follows an S-shaped curve.

Various threshold theories propose different explanations for the lack of a fixed threshold (Corso 1963). In modern psychophysics, the signal detection theory has exerted a profound influence on the concept of a sensory threshold (Swets 1961; Green and Swets 1966). According to this theory there exists, within sensory systems, a substantial amount of noise activity. The noise is added to the true sensory signal elicited by a peripheral stimulus before the signal is read off. As a result, it is impossible for a human observer to detect very small afferent signals altogether, or rather, he is not able to discriminate between noise alone and a small sensory signal with noise added. Larger inputs would be detected with successively greater probability until a probability of one is attained with suprathreshold stimuli.

It is assumed that a human observer, when facing a detection task, defines for himself a decision criterion in terms of signal size within the sensory system. When the size of the signal exceeds this level, the subject reports a sensation, regardless of a stimulus having been delivered or not, i.e. regardless of whether the signal which he is reading off is accounted for by noise alone or is a true sensory signal plus noise. Obviously, he has no means to discriminate between the two. Thus the position of the threshold curve along the stimulus strength axis would partly be determined by the subject's decision criterion (Green and Swets 1966).

[1] This study was supported by the Swedish Medical Research Council (Grant 14X–3548), Umeå University (Fonden för Medicinsk Forskning) and Gunvor and Josef Anérs Stiftelse.

It may be speculated that the neurophysiological basis of the noise is a continuous and spontaneous activity in neurons of the sensory system, although it should be pointed out that noise assumption is based on psychophysical rather than neurophysiological data (Green and Swets 1966).

Obviously, a key element in the signal detection theory is the assumption that the sensory threshold is largely set by central mechanisms, and that the psychophysical threshold is elevated above the detection capacity of the peripheral sense organs. The detection probability would largely be determined by the amount of noise within the sensory system and by the subject's decision criterion. In fact, many psychophysical studies have been devoted to analyses of the characteristics of these mechanisms.

An alternative view which has been supported by neurophysiological studies is that the psychophysical threshold is set by the sensitivity of the receptor cells in the periphery rather than by central mechanisms within the brain and spinal cord. Hecht et al. (1942) provided arguments for the absolute threshold of the eye being set by the properties of the rods. They demonstrated that a sensation of light was reported when only one light quantum per rod was absorbed by 5–14 rods. A different type of threshold was studied by Mountcastle and coworkers, who found that the threshold for a sensation of vibration was set by the receptors in the glabrous skin (Talbot et al. 1968; Mountcastle et al. 1969). In both sets of studies, the sensation was dependent on the simultaneous activation of a number of afferents.

This chapter, which is largely based on previously published material, examines the threshold concept in relation to tactile stimuli in the human hand. Evidence will be presented that the absolute threshold for touch stimuli may be set by the properties of the receptors in the glabrous skin. Moreover, it was found that summation was not required but that a single impulse in a single Meissner unit may produce a secure detection of the stimulus.

Methods

Nerve impulses from single afferent units were recorded with the microneurographic method (Vallbo and Hagbarth 1968; Vallbo 1972). A tungsten needle electrode was inserted through the intact skin of attending human subjects into the median nerve on the upper arm. In the experiments of the present study, recording was limited to tactile afferent units.

In sessions of microstimulation, the recording electrode was connected to a stimulator and constant current pulses were delivered in order to stimulate a single afferent fibre (Vallbo 1981; Vallbo et al. 1984). The psychophysical threshold was determined with the method of constant stimuli, using either the yes/no or the two alternative, forced choice procedure (Gescheider 1976; Johansson and Vallbo 1979).

Mechanical stimuli were delivered with a small probe which had a rounded tip with a diameter of 0.45 mm. Considerable precautions were taken to produce indentations of accurately defined amplitudes, in spite of the continuous movements of the skin surface which inevitably occur with pulse and respiratory activity even when the hand is rigidly fixed. To achieve accurate indentations, a stimulator was developed which operated in relation to the skin surface rather than in relation to a rigid external support (Westling et al. 1976; Johansson, and Vallbo 1979). Exact localisation of the stimulus was ensured with a continuous control of the target area through a microscope. Small movements of the subject's hand were compensated manually when necessary. The stimulation method was developed with the aim to activate one single afferent unit in isolation. It has actually been shown in a separate study that it is adequate to activate a single Meissner corpuscle of a fast adapting type I (FA I) unit (Johansson 1978).

Care was taken to avoid clamps and bands for fixation in order to eliminate distracting sensory stimuli of the glabrous skin. The purpose was to deliver the test stimuli against a silent background in order to provide optimal conditions for detection of minimal sensory inputs. Stimuli were triangular indentations with a probe speed of 4 mm per second.

Results

Microneurography recordings were attained by careful adjustment of the electrode position within the nerve until, by chance, impulses from one single unit could be discriminated from noise and discharge in other fibres. Examples are shown in Fig. 13.1, which displays responses of two different kinds of tactile units in the glabrous skin areas. Although the noise level is always relatively high in microneurography recordings (10–20 μV peak-to-peak), there is no difficulty in discriminating unitary impulses in good recordings. A single unit may occasionally last for 1–3 h, whereas most of the data were extracted from recordings lasting 10–20 min.

Unit Types in the Glabrous Skin of the Human Hand

It has been demonstrated that there are four different kinds of tactile units in the glabrous skin of the human hand (Fig. 13.2). Four similar types have previously been described in other species, although the slowly adapting type II (SA II) unit has not been found, except occasionally, in the glabrous skin of subhuman primates.

The fast adapting units, FA I and FA II, have previously been denoted RA or QA and PC. It is suggested that these terms should be eliminated because they are inadequate and often confusing (Vallbo 1984). For instance, when the term "RA" is used to denote central neurons in the somatosensory system, it is often not clear whether the term denotes any neuron which is rapidly adapting to the particular stimuli delivered or whether particular afferent connections are implied.

Psychophysical Thresholds to Touch Stimuli

When the psychophysical threshold was measured, it was found that there were consistent regional differences between separate skin regions of the hand with

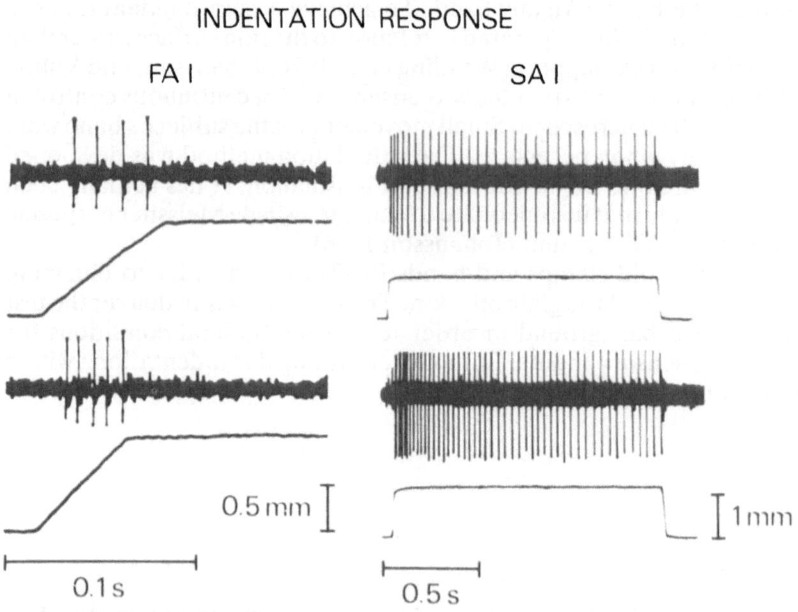

Fig. 13.1. Sample records of microneurographic recordings from single tactile afferents. Recordings were obtained with a needle electrode inserted percutaneously in the median nerve 8 cm above the elbow. Responses are shown of two different types of tactile afferent units to localised skin indentations of the glabrous skin, i.e. a fast adapting type I unit (*FA I*), which probably is connected to Meissner corpuscles, and a slowly adapting type I unit (*SA I*), which is connected to Merkel cells.

		RECEPTIVE FIELDS	
		Small, sharp borders	Large, obscure borders
ADAPTATION	Fast, no static response	FAI (RA) Meissner	FAII (PC) Pacini and Golgi-Mazzoni
	Slow, static response present	SAI Merkel	SAII Ruffini

Fig. 13.2. Types of tactile afferent units found in the glabrous skin of the human hand. Some of the distinguishing properties, the notations and the type of end organs in the skin are indicated.

regard to subjects' ability to detect weak stimuli. This is illustrated in Fig. 13.3, which shows group data from 51 subjects. The psychophysical thresholds were around 10 μm of indentions for the major part of the glabrous skin. However, in the centre of the palm and in some other regions, which are specified in the figure legend, the thresholds were considerably higher and the variability was much larger. Thus, it seemed justified to consider two main regions in the glabrous skin of the human hand, i.e. a high-threshold region and a low-threshold region. Although this unexpected finding complicated the analysis, it also pointed towards interesting relationships between neural and perceptive thresholds, as will be considered below.

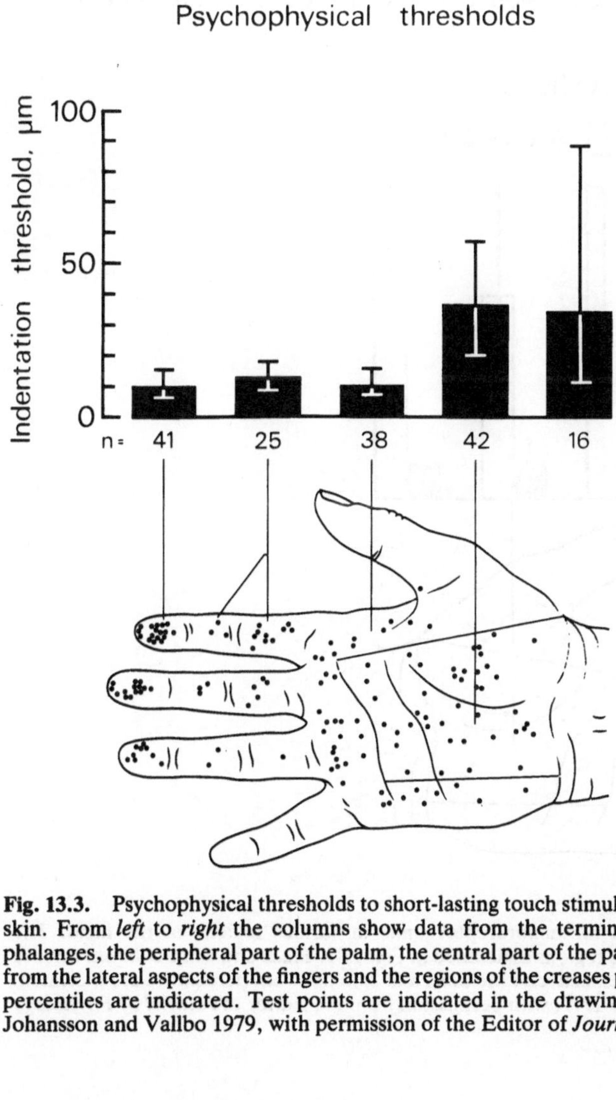

Fig. 13.3. Psychophysical thresholds to short-lasting touch stimuli in various regions of the glabrous skin. From *left* to *right* the columns show data from the terminal phalanx, the middle and basal phalanges, the peripheral part of the palm, the central part of the palm, and, to the extreme *right*, data from the lateral aspects of the fingers and the regions of the creases pooled. Medians and 25th and 75th percentiles are indicated. Test points are indicated in the drawing of the hand. [Reproduced from Johansson and Vallbo 1979, with permission of the Editor of *Journal of Physiology* (London)]

Thresholds of Primary Afferents to Touch Stimuli

The thresholds of the afferent units fell into two groups, as shown in Fig. 13.4. The FA units had thresholds around 10 μm, whereas the SA units exhibited much higher thresholds. Data were pooled from the two types of fast units as well as from the SA units because no interesting differences were found between type I and type II units in this respect.

Relationship Between Psychophysical and Afferent Thresholds

A comparison between the two sets of data presented in Figs. 13.3 and 13.4 clearly shows that the thresholds of the SA afferent units were considerably higher than the psychophysical thresholds in the low-threshold region. This finding allows the definite conclusion that the SA units cannot account for

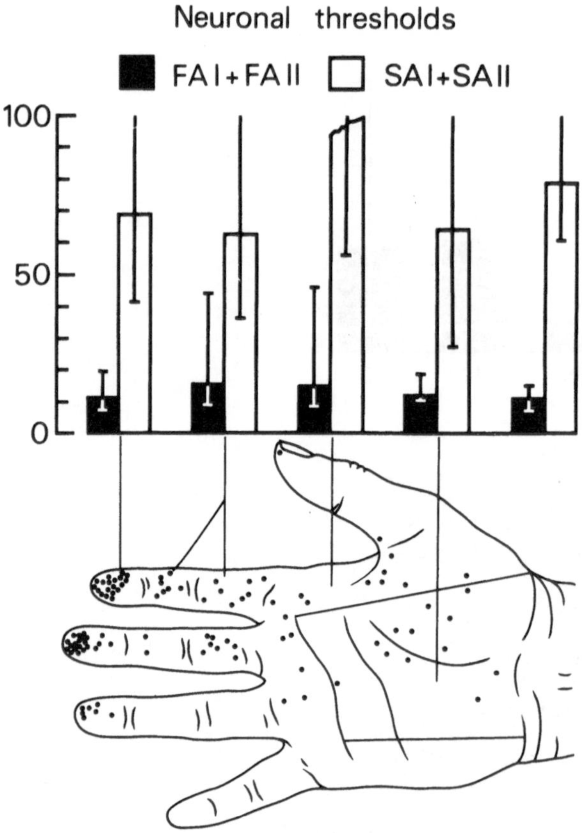

Fig. 13.4. Thresholds of afferent mechanoreceptor units in various regions of the hand. The successive columns refer to the same regions as detailed in Fig. 13.3. *Black* and *white columns* refer to rapidly and slowly adapting units respectively. Medians and 25th and 75th percentiles are indicated. [Reproduced Johansson and Vallbo 1979, with permission from the Editor of *Journal of Physiology* (London)].

psychophysical detection in the major part of the glabrous skin region, simply because the SA units did not respond to these stimuli. Thus psychophysical detection must be dependent on signals from the FA units. This is not to deny that impulses in SA units may account for a sensation. Actually, it will be shown below that even a single SA I unit may give rise to a sensation of light touch.

A second point which emerges from a comparison between Figs. 13.3 and 13.4 is that the thresholds of the FA units were very similar to the psychophysical thresholds. In particular, it deserves emphasis that their thresholds were not appreciably lower. This indicates that stimuli which are readily detected by human observers produce only a very small afferent signal. It can be inferred that it consists of one action potential either in a small group of afferents or even in one single afferent unit. The exact number of units which are activated at threshold will be considered below.

Although it is obvious from Fig. 13.3 that there were pronounced differences in psychophysical thresholds between different skin regions, it is equally clear from Fig. 13.4 that there are no corresponding differences of the afferents' thresholds. This indicates that the higher psychophysical thresholds in the centre of the palm are not due to the sense organs being less sensitive here. One possibility is that the density of units is lower. An alternative is that the afferent signal is handled differently in the spinal cord and the brain, i.e. that mechanisms within the central nervous system rather than peripheral factors account for the regional threshold differences in the hand. This problem will be considered below.

Figure 13.5 summarises schematically the data on psychophysical and afferent

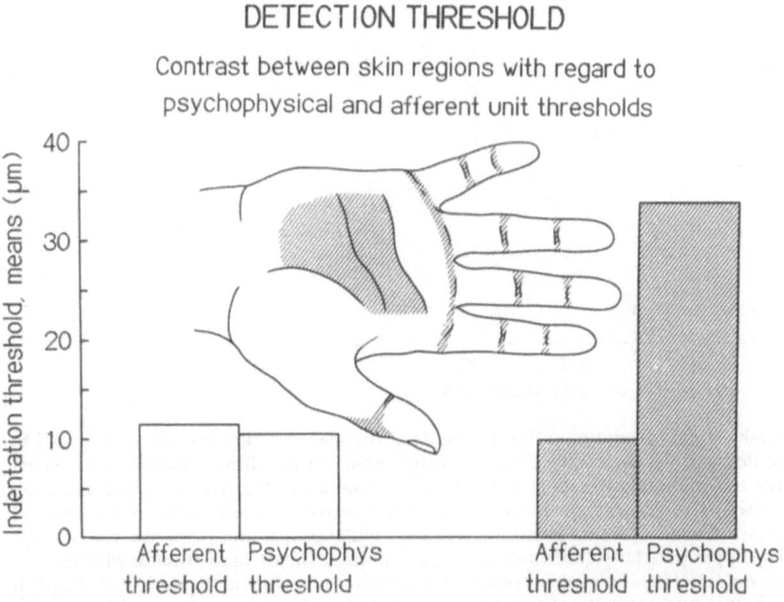

Fig. 13.5. Schematic presentation of high- and low-threshold regions of the glabrous skin of the human hand.

thresholds in the hand to highlight the contrast between two skin regions. The stippled areas indicate regions of high psychophysical thresholds, which include also the lateral aspects of the finger.

Threshold Curves

The findings presented in Figs. 13.3 and 13.4 are based on experimental data of the kind illustrated in Fig. 13.6. It shows threshold data extracted with the method of constant stimuli. A number of stimuli of varying amplitudes were delivered and two kinds of responses were assessed. One was the percentage of "yes" responses in yes/no tests or the percentage of correct choices in two alternative forced choice tests, which then were plotted as a function of stimulus

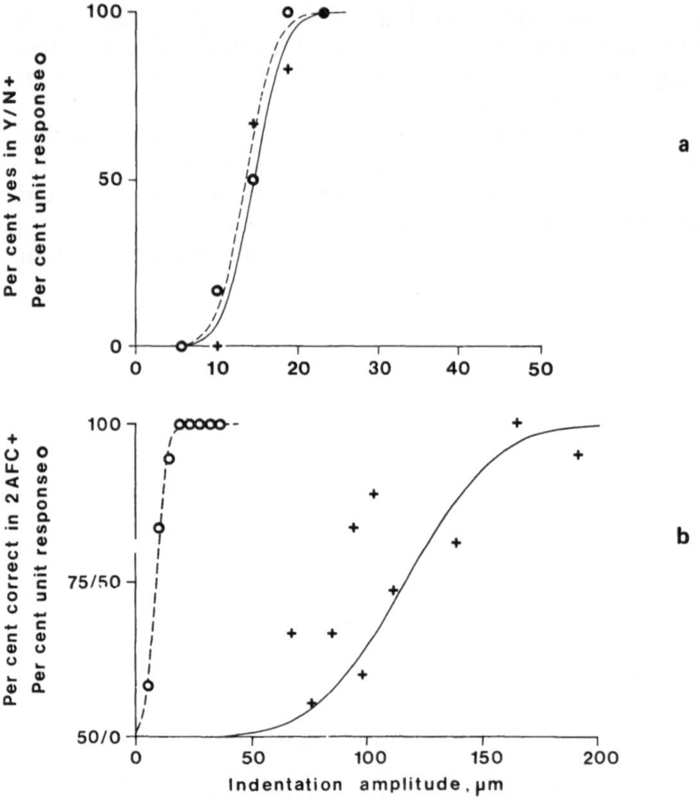

Fig. 13.6a,b. Threshold data from two different points in the glabrous skin are shown in **a** and **b**. Touch stimuli were delivered at the centre of the receptive field of two different tactile units, which were recorded with the microneurographic method. Simultaneously, the psychological detection responses were assessed. The *crosses* represent percentage of positive responses in the psychophysical tests, whereas the *open circles* represent percentage of tests with positive unit response, i.e. a single action potential. The upper panel (**a**) shows data from a test point at the tip of the thumb with an FA I unit, whereas the lower panel (**b**) shows corresponding data from a test point in the middle of the palm while another FA I unit was recorded. Cumulative normal probability curves were fitted to the experimental data by a method of maximal likelihood estimation. (Reproduced from Vallbo and Johansson 1976, with permission of the publishers, Pergamon Press, Oxford)

amplitude. A large number of false trials were intermingled with the real stimuli in the yes/no procedure. In addition, the percentage of tests in which the afferent unit responded with one impulse or none at all was assessed and plotted in the same diagrams.

The two panels of Fig. 13.6 refer to two different points in the glabrous skin, as detailed in the legend. When the neural and psychophysical thresholds were compared at the same point of stimulation it was found that these were sometimes identical for Meissner corpuscle units, as in Fig. 13.6a. The very close agreement between the two threshold curves suggests that the particular unit recorded was actually the one which provided the crucial signal for detection. Moreover, considering that the method of stimulation was accurate enough to activate an individual Meissner corpuscle within the receptive field (Johansson 1978), it seems reasonable that these stimuli excited only the one single unit which was recorded.

The data of Fig. 13.6b were collected from the centre of the palm. Here, the psychophysical threshold was much higher than the afferent's threshold. The high psychophysical threshold was not due to a lack of sensory signals because the microneurography recording revealed that the unit responded with a number of impulses to many of the stimuli which the subject failed to detect. Moreover, although the recording was limited to one afferent unit, indirect evidence indicated that additional afferents were activated as well, by many of the stimuli (Johansson and Vallbo 1976). Hence, both a spatial and a temporal summation were required to produce a sensation of touch in the central part of the palm. It is also evident from Fig. 13.6b that the psychophysical data were much more scattered and the slope of the curve was much lower. These findings are all consistent with the interpretation that a considerable noise is added to the afferent signal before it is read off by the subject.

Psychophysical and Afferent Responses in Individual Trials

The interpretation suggested by Fig. 13.6a that subjects were able to detect, in psychophysical tests, stimuli when only one single unit was activated was further corroborated in analyses of neural and psychophysical responses of individual trials, as illustrated in Fig. 13.7. The lower two panels of this figure represent 30 successive tests in which the stimulus lingered exactly on the threshold of the afferent unit. It may be seen that in some tests a single impulse was evoked, as indicated by the triangle at the lower end of the bar, whereas no discharge appeared in others. The subject's "yes" response is indicated by a triangle at the upper end of the bar. It may be seen that the coincidence of a spike in the afferent nerve fibre and the subject's "yes" response was almost exact. Only in one single test did the subject fail to report a sensation when a nerve impulse was elicited. This coincidence of the data strongly suggests that the single nerve impulse from this unit, in fact, accounted for the psychophysical detection of the touch stimuli.

Population Response

It may be argued that another afferent unit might have been activated in addition to the one recorded in Fig. 13.7, implying that detection of touch stimuli would be

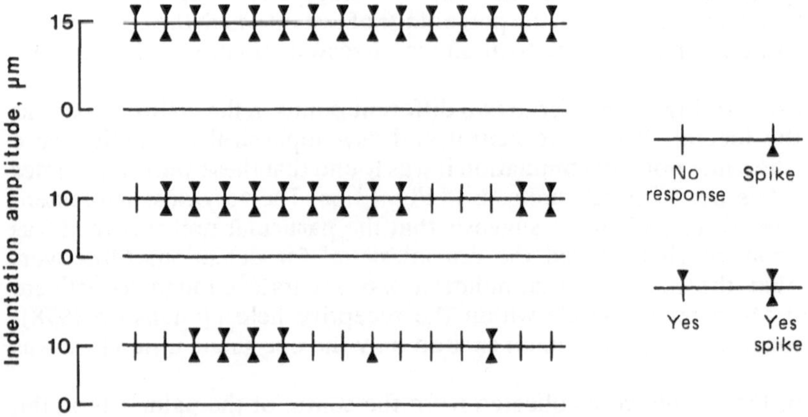

Fig. 13.7. Neural and psychophysical responses of individual touch stimuli. Forty-five stimuli were delivered in succession at the most sensitive point of an FA I unit located at the distal phalanx of the index. The *short vertical bars* represent the stimuli, whereas neural and psychophysical responses are indicated by *filled triangles*, as shown by the key to the *right*. Indentation amplitudes are given by the ordinate values corresponding to the horizontal lines. It is obvious that an almost perfect agreement between unit response and the subject's sensation of a touch stimulus was present. (Reproduced from Vallbo and Johansson 1976, with permission of the publishers, Pergamon Press, Oxford)

dependent on a larger sensory signal, i.e. one impulse in two or several units. This alternative interpretation was explored using two different approaches, as will be considered below, i.e. an estimation of population response and microstimulation through the needle electrode.

In order to explore further the probability that one or several units were activated by the critical touch stimuli delivered in the detection analysis, population responses were inferred. The number of tactile units which innervate the various skin regions of the hand were estimated on the basis of two sets of data. One set was the relative number of units sampled from the separate regions and the other was the total number of nerve fibres within the median nerve. These data were combined with receptive field data and unit threshold data and constituted a basis for an estimation of the number of tactile units which would be activated as a function of stimulus intensity (Johansson and Vallbo 1976; Johansson and Vallbo 1979). The estimate clearly indicated that only one afferent unit was excited at psychophysical detection threshold in a large proportion of tests (Fig. 13.8). Hence independent findings strongly supported the conclusion suggested by Figs. 13.6a and 13.7 that the detection of a touch stimulus may be based on one single impulse in a single Meissner unit in large regions of the glabrous skin.

Microstimulation

Still another independent approach was applied to assess the size of the minimal afferent signal required for the production of a sensation. This approach was based on the method of microstimulation of identified afferent units (Torebjörk and Ochoa 1980; Vallbo 1981; Ochoa and Torebjörk 1983; Vallbo et al. 1984). In these experiments, a tactile unit in the peripheral nerve was first recorded,

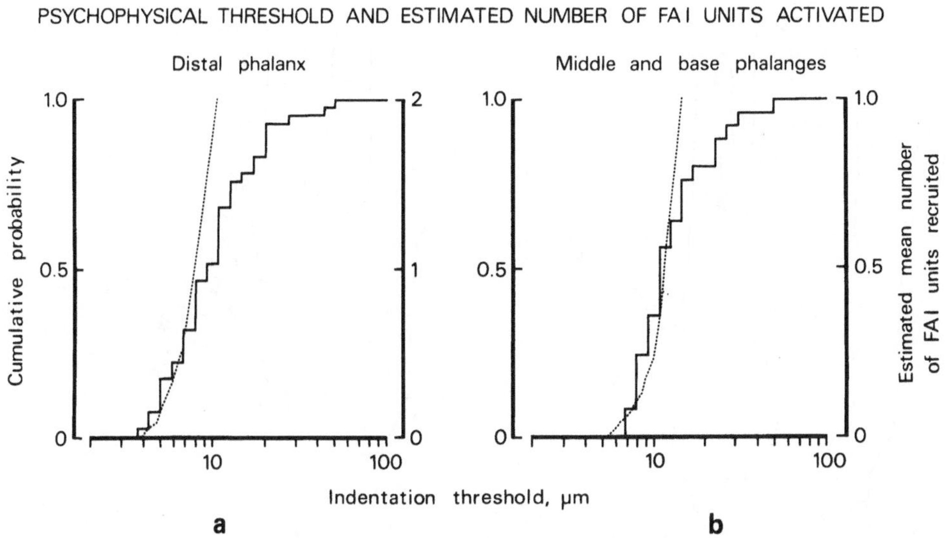

Fig. 13.8a,b. Comparison between psychophysical thresholds and estimated number of responding FA I units as a function of indentation amplitude in two different skin regions, i.e. the terminal phalanx (**a**) and the middle and the base phalanges (**b**). The *continuity lines* represent, in cumulative form (*left ordinates*), the distribution of the psychophysical thresholds of the whole sample, whereas the *interrupted lines* represent the estimated average number of FA I units activated as a function of indentation amplitude. The psychophysical threshold was defined as the mid-point of threshold curves, as illustrated in Fig. 13.6.

It may be seen that, on the average, not more than one FA I unit, responded at 10 μm on the finger tip, whereas on the middle and base phalanges 0.5 unit responded at this stimulus intensity, which was a common threshold value in the psychophysical tests. [Reproduced from Johansson and Vallbo 1979, with permission of the Editor of *Journal of Physiology* (London)].

identified, and characterised. The recording electrode was then reconnected to an electrical stimulator and a train of pulses of suitable amplitude were injected, while the subject's perceptive experience of the stimulus was explored with psychophysical methods. A number of findings indicate that one type of sensation elicited by microstimulation depended on the activation of one single unit alone, and often the particular unit which was recorded immediately before the stimulation (Ochoa and Torebjörk 1983; Vallbo et al. 1984).

When a train of pulses with an amplitude of about 1 μA were delivered through the recording electrode, a distinct and clear sensation was reported in typical cases. It was felt within a small skin area: only a couple of square millimetres. It was very clear and well defined in several respects, e.g. with regard to location and extent on the skin surface. The quality of the sensation was practically always of a mechanical nature. The intensity was described as faint but clearly above threshold for detection.

These well-localised sensations had a quantal nature in that they appeared at a definite current strength (median 0.81 μA). A further increase of stimulus intensity above the level required for a single sensation resulted in one of several additional sensations of similar kind located at semi-random distance from the first one, as if successive afferent fibres were recruited (Fig. 13.9). With higher current intensities, the subjects reported sensations of paraesthetic nature which covered much larger skin areas. The size of the skin area where an individual

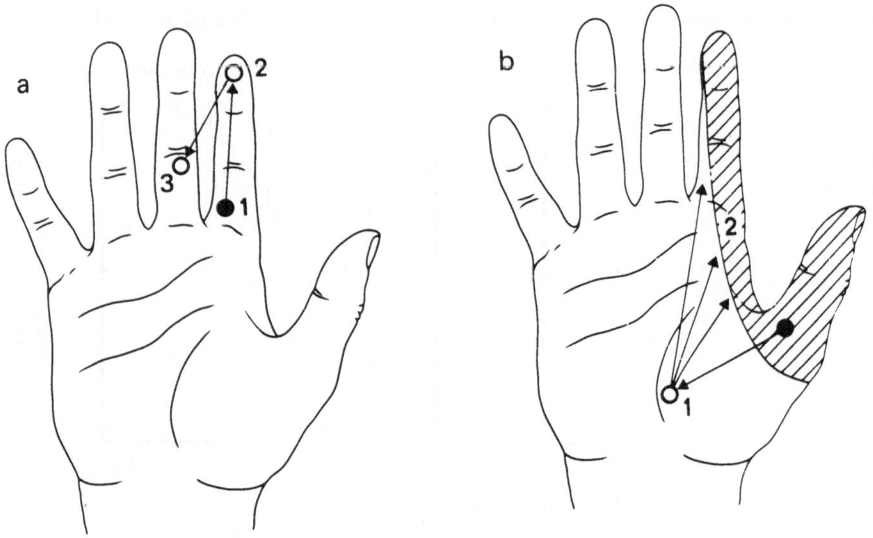

Fig. 13.9a,b. Quantal nature of sensations elicited by intraneural microstimulation. The drawings show the location of sensations reported in response to train stimuli of successively increasing current intensity. In **a** a localised sensation was reported exactly on the receptive field of an FA I unit, recorded immediately before (*filled circle*). When current intensity was increased, two additional sensations of similar nature were successively reported (*open circles*) indicating the recruitment of two additional single nerve fibres. In **b** the first sensation reported by the subject on successively increasing stimulus intensity was at a point remote from the receptive field of the unit (*1*). A further increase of stimulus intensity produced a paraesthetic sensation in a large area (*2*), indicating activation of a large number of units. (Reproduced from Vallbo et al. 1984, with permission of the Editor of *Brain*).

sensation was felt was usually similar to the size of the receptive field of the FA I and SA I units.

When an afferent unit had been recorded immediately beforehand, the properties of the sensations usually matched the properties of the unit. For instance, the quality of sensation was dependent on unit type. In the example of Fig. 13.10, an SA I unit with a receptive field close to the nail of the index was first recorded. When microstimulation was then applied, the subject reported a sensation of uniform, sustained pressure within a small area covering the receptive field of the afferent, as described in the legend. With FA I and FA II units, on the other hand, the sensation was usually of a different quality in that it had a clear, vibrating, wobbling or buzzing character. The matching of quality to unit type comes out more clearly with fairly long pulse trains but was not as clear with short trains (Ochoa and Torebjörk 1983; Vallbo et al. 1984).

The matching between unit properties and sensation was often very striking in respect of spatial aspects. The similarity of the afferent's receptive field and the extent on the skin of the sensation was often remarkably accurate. For instance, with the unit of Fig. 13.10 the size, shape and even the orientation of the small skin area where the sensation was felt was practically identical to the receptive field of the unit recorded immediately beforehand. It is difficult to avoid the conclusion that the sensation was based on impulses in this afferent unit.

When the microstimulation approach was used to assess the minimal afferent input required for a sensation, the procedure outlined above was first pursued, i.e. a unit was located for study and its functional properties were determined as well

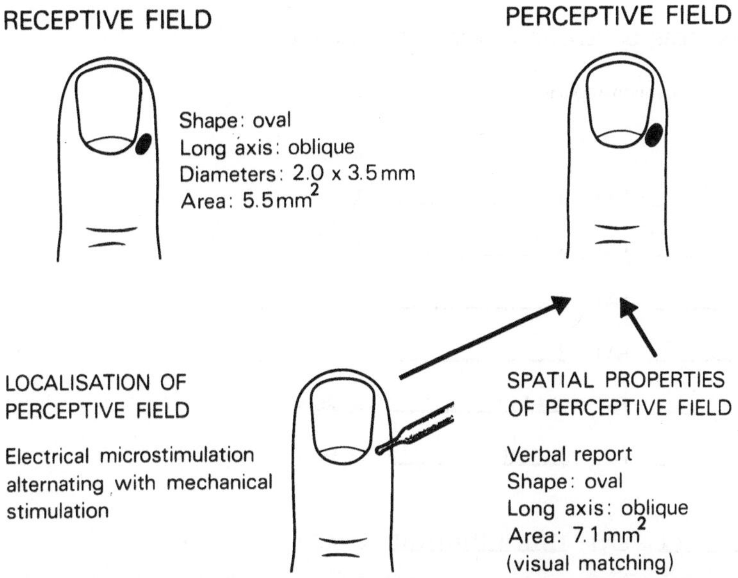

Fig. 13.10. Identity of unitary receptive field and extent of sensation experienced by the subject on microstimulation. The receptive field of an SA I unit was first assessed (*upper left panel*) using stimuli at four times threshold. When microstimulation was applied immediately afterwards the subject reported a sensation of light continuous indentation in the same skin region. In order to assess its exact location, mechanical stimuli and electrical pulse trains were delivered alternatively (*lower panel*). The subject then experienced two touch stimuli occurring shortly after each other in time. His task was to define whether they were present at the same spot or not. The subject was then asked to define size, shape and orientation of the skin area where the sensation was felt (*upper right panel*).

as the properties of the sensation associated with train stimuli. Then the minimal number of stimulating pulses was defined which elicited a just noticeable sensation. An outcome of a number of such tests is illustrated in Fig. 13.11. It may be seen that when a Meissner (FA I) unit was stimulated one single pulse was often sufficient to produce a sensation of touch. Obviously, this finding confirms the conclusion based on the previous analysis using mechanical stimuli.

On the other hand, it was not a general finding that one single nerve impulse in a single fibre produced a sensation. First, it was not true for all the Meissner (FA I) units, as shown in Fig. 13.11. It remains to be defined whether those FA I units which are effective in this respect have some common properties. It may be speculated that the location of the receptive field is an essential factor, but the published material is not large enough to allow any conclusion in this respect. Second, it was found that the stimulation of a single SA I unit was not associated with a sensation of touch unless a fairly large number of impulses were initiated.

To summarise, the microstimulation experiments indicate that a single impulse from one of the FA I units constitutes a sensory signal which may penetrate to consciousness and produce a sensation of touch. It might be relevant that the most sensitive end organs to the touch stimuli used in the present study are found in this group of units. The findings also demonstrate that impulses in single SA I units may produce a sensation of touch. However, with these units a large number of impulses are regularly required, implying that temporal summation is necess-

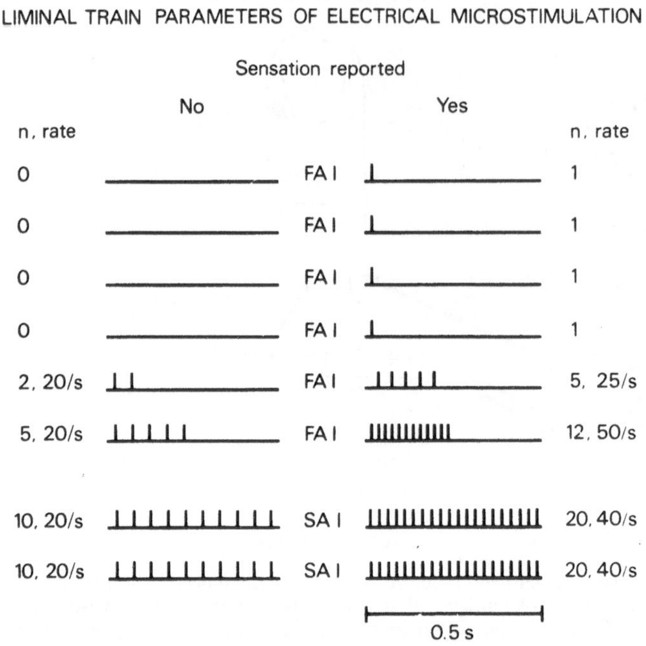

Fig. 13.11. Liminal train parameters for eliciting a sensation of touch on intraneural microstimulation of two different kinds of units. Immediately before stimulation a single unit was recorded of the type indicated (FA I and SA I), and it had been concluded that the sensation was due to the activation of this unit. The receptive fields were all located on the volar aspects of the fingers except for the last FA I unit in the column, whose field was located on the dorsal aspect of the index. It may be seen that a single impulse elicited a sensation of touch when many of the Meissner units were stimulated. (Reproduced from Vallbo et al. 1984, with permission of the Editor of *Brain*).

ary in this system when a single fibre is activated. It might be of significance that the SA I units have invariably higher thresholds to touch stimuli than the FA I units, as shown in Fig. 13.4.

Discussion

Two independent analyses agree that a single action potential in a single Meissner unit may produce a detectable sensation of touch within the human mind. One is the finding with microneurography recording that psychophysical detection of mechanical touch stimuli occurs when a single impulse is initiated in one of the most sensitive afferents. The other is the demonstration that electrical stimulation of single FA I units through the microneurography electrode was associated with a sensation of touch when not more than one impulse was initiated in the afferent.

The findings demonstrate that summation at the level of the peripheral afferents is not necessary to produce a sensation of touch. That temporal summation is not necessary is evident from the finding that only one impulse was elicited in the afferent when psychophysical detection regularly occurred. Spatial summation, on the other hand, was not required, which was evident from the finding that the two methods of stimulation excited only one single afferent at threshold of detection. These conclusions clash with one of the main points in the view advocated by Wall and coworkers, who claim that "more than one fibre carrying single nerve impulses is essential for central cells to detect the characteristics of a sensory stimulus" (Melzack and Wall 1962; Wall and McMahon 1985; Torebjörk et al. 1987).

The finding that one action potential in a single nerve fibre may produce a sensation of touch implies that sensory threshold may, in fact, be a step function, when the probability of detection is considered in relation to the afferent signal rather than stimulus strength. It is a corollary that the variability of the psychophysical response to stimuli of nominally constant strength was due to peripheral and not to central factors. It was accounted for by a variability of the effective stimulus seen by the sense organ or a variability of the receptor responsiveness, rather than central mechanisms such as noise in the spinal cord or the brain as assumed in the signal detection theory. This conclusion is in agreement with the finding of Hecht et al. (1942) that stimulus variability accounted for the psychophysical detection curve in vision.

The findings imply that the signal detection theory is not relevant for large regions of the glabrous skin when touch stimuli are studied. The data when nerve impulses and psychophysical responses were recorded simultaneously (see Fig. 13.7) have a particular relevance in this context. If nerve impulses had not been recorded in this experiment, it would have been reasonable to interpret the variation in psychophysical response as an effect of noise within the central nervous system, considering that the series of stimuli were of identical amplitude. However, the simultaneous recording of nerve impulses clearly demonstrated that the stimuli which the subject failed to detect were, in fact, ineffective in the sense that they did not excite the afferent unit. They were equivalent to false trials. Hence, the test series of Fig. 13.7 represents a regular yes/no procedure with about equal number of true stimuli and catch trials. The fact that the subject did not respond to any of the stimuli which failed to elicit a nerve impulse implies that his false alarm rate was zero. It is highly significant for the theoretical interpretation that the hit rate is practically 100% at the same time as the false alarm rate is 0%.

In the signal detection paradigm, the false alarm rate as well as the hit rate is manipulated with a reward system in order to demonstrate that the subjects' decision criterion can be shifted to lower or higher levels. The theory predicts that, by increasing the rate of false alarms, the rate of correct "yes" responses would also increase. With the data of Fig. 13.7 it would be totally pointless to force the subject to produce a higher rate of false alarms because this would not change his hit rate. It was also experimentally found that attempts to produce more false alarms failed with practically all subjects.

The combination of 100% hits with 0% false alarms clearly indicates that the signal detection theory is not applicable for stimuli delivered in the low-threshold region of the human hand. With data of this nature an ROC analysis is inadequate because the data do not conform with the basic assumption behind this analysis.

Or, to put it in a different way, the single action potential represents a sensory signal which is significant enough to stand out clearly from any noise there may be in the somatosensory system. This, in turn, implies not only that the threshold is a step function but also that that the theoretical minimal sensory signal is securely detected. The signal detection theory, on the other hand, assumes that a minimal sensory input cannot be securely detected.

In the introduction to this chapter, the question was considered whether the psychophysical threshold is set by the sensitivity of the peripheral sense organs or by central mechanisms. The present findings suggest that both may be true. In large regions of the glabrous skin of the human hand the absolute touch threshold is set by the sensitivity of the most sensitive peripheral sense organs. However, this is not generally true, but it is valid for the finger pads and the peripheral part of the palm. For other regions in the hand, e.g. the central part of the palm, the limit is apparently set by central mechanisms within the brain.

Hence, the present data indicate that the sensory threshold is not uniquely set by one factor but that both main alternatives might be true for different parts of the somatosensory system. It seems that the threshold is set by the peripheral sense organs for the skin regions which are of greatest importance in exploration and fine manipulation, e.g. the finger pads, which have been called the macula of the tactile system. This implies that the analysis of the afferent signal within the brain involves the secure discrimination between no impulse and a single impulse in one afferent fibre of the nerve. It may be speculated that for other skin regions which probably have a less prominent role in tactile functions less analytical capacity has been reserved within the brain.

The mechanism accounting for the observation that the afferent signal is handled differently in separate sections of the somatosensory system is not clear. In psychophysical terms, it may be due to a difference in noise level within central parts of the somatosensory system. A compatible interpretation is that the crucial factor is the amount of neuronal activity in the somatosensory cortex evoked by a small afferent input. Whether a postulated variation of this nature is related to a variation in the relative size of cortical representation is an open question, although such a dependence seems plausible.

References

Corso JF (1963) A theoretico-historical review of the threshold concept. Psychol Bull 60:356–370

Gescheider GA (1976) Psychophysics. Method and theory. Wiley, New York

Green DM, Swets JA (1966) Signal detection theory and psychophysics. Wiley, New York

Hecht S, Slaher S, Pirenne MH (1942) Energy, quanta and vision. J Gen Physiol 25:819–840

Johansson RS (1978) Tactile sensibility in the human hand: Receptive field characteristics of mechanoreceptive units in the glabrous skin area. J Physiol (Lond) 281:101–123

Johansson RS, Vallbo ÅB (1976) Skin mechanoreceptors in the human hand: an inference of some population properties. In: Zotterman Y (ed) Sensory functions of the skin in primates. Pergamon, Oxford, pp 171–184

Johansson RS, Vallbo ÅB (1979) Detection of tactile stimuli. Thresholds of afferent units related to psychophysical thresholds in the human hand. J Physiol (Lond) 297:405–422

Melzack R, Wall PD (1962) On the nature of cutaneous sensory mechanisms. Brain 85:331–356

Mountcastle VB, Talbot WH, Sakata H, Hyvarinen J (1969) Cortical neuronal mechanisms in flutter-vibration studied in unanesthetized monkeys. Neuronal periodicity and frequency discrimination. J Neurophysiol 32:452–484

Ochoa J, Torebjörk HE (1983) Sensations evoked by intraneural microstimulation of single mechanoreceptor units innervating the human hand. J Physiol (Lond) 342:633–654

Swets JA (1961) Is there a sensory threshold? Science 134:168–177

Talbot WH, Darian-Smith I, Kornhuber HH, Mountcastle VB (1968) The sense of flutter-vibration: comparison of the human capacity with response patterns of mechanoreceptive afferents from the monkey hand. J Neurophysiol 31:301–334

Torebjörk HE, Ochoa J (1980) Specific sensations evoked by activity in single identified sensory units in man. Acta Physiol Scand 110:443–447

Torebjörk HE, Vallbo ÅB, Ochoa JL (1987) Intraneural microstimulation in man: its relation to specificity of tactile sensations. Brain 110:1509–1529

Vallbo ÅB (1972) Single unit recording from human peripheral nerves: muscle receptor discharge in resting muscles and during voluntary contractions. In: Somjen GG (ed) Neurophysiology studied in man. Excerpta Medica, Amsterdam, pp 283–297

Vallbo ÅB (1981) Sensations evoked from the glabrous skin of the human hand by electrical stimulation of unitary mechanosensitive afferents. Brain Res 215:359–363

Vallbo ÅB (1984) Tactile sensation related to activity in primary afferents with special reference to detection problems. In: von Euler C, Franzen O, Lindblom U, Ottoson D (eds) Somatosensory mechanisms. Macmillan, London, pp 163–172

Vallbo ÅB, Hagbarth K-E (1968) Activity from skin mechanoreceptors recorded percutaneously in awake human subjects. Exp Neurol 21:270–289

Vallbo ÅB, Johansson RS (1976) Skin mechanoreceptors in the human hand: neural and psychophysical thresholds. In: Zotterman Y (ed) Sensory functions of the skin in primate. Pergamon, Oxford, pp 185–199

Vallbo ÅB, Olsson KA, Westberg K-G, Clark FJ (1984) Microstimulation of single tactile afferents from the human hand: sensory attributers related to unit type and properties of receptive fields. Brain 107:727–749

Wall PD, McMahon SB (1985) Microneuronography and its relation to perceived sensation. A critical review. Pain 21:209–229

Westling G, Johansson RS, Vallbo ÅB (1976) A method for mechanical stimulation of skin receptors. In: Zotterman Y (ed) Sensory functions of the skin in primates. Pergamon, Oxford, pp 151–158

Section V
MOTOR SYSTEM

Chapter 14

Order and Disorder in the Motor System

M. Swash

Introduction

The problem of motor control is central to the contemporary neurosciences, especially to neurology, since it involves an understanding of the basic mechanisms of the function of the nervous system, and its relation to its environment. The brain, it is said, exists to provide effector interaction with the external world. The famous dictum of Aristotle: "Nihil est in intellectu quod non primus fuerit in sensu" serves to indicate the essentially subjective nature of enquiry into brain function, associated in European minds with the philosophy of Descartes, which was the received approach before the advent of the experimental method. It serves also, perhaps, to remind us how constrained even current research is in humans, despite the availability of newer methods of functional imaging and measurement. Ferrier (1876) regarded the human brain as possessing both sentient and active or motor functions, and did not clearly relate these to each other despite his insights into the role of the cerebral cortex in the initiation of movement. The stimulus for continued investigation of the problems posed in seeking understanding of motor control has largely derived from the need to seek alleviation of the disabilities suffered by patients with diseases of the brain. These disabilities have proven difficult to classify and to describe. Furthermore, the fragmentation of investigation that resulted from the separation of the neurosciences from clinical research has led to further confusion in understanding these problems. This separation was apparent relatively early in the development of concepts of motor control.

Hughlings Jackson (Taylor 1931/32), to whose contributions this book is dedicated, worked in relatively complete isolation in clinical neurological practice, developing his ideas of the integration and dissolution of nervous function from observations made on patients afflicted with a variety of different and ill-understood disorders of the nervous system. His underlying hierarchical concept of brain function was unique, being derived from the evolutionalist ideas of Herbert Spencer, and relatively uninfluenced by the other conventional doctrines of his day. Jackson's ideas themselves were taken up abroad, particularly in other European countries, and they continued to influence generations of medical students, and especially students of clinical neurology. Their significance for neuroscience was less appreciated because of the relative dominance of the Sherringtonian school in neurophysiology and the later emphasis on microelectrode technology in the investigation of brain function (for example, see Stein 1986). Yet clinical observations of disordered motor control, coupled with precise measurement of the patterns of functional disorder, still have much to teach us.

The precise and rapid ability to make movements with a limb, or with part of a limb; to follow moving stimuli with the eyes or with a hand; to make complex postural adjustments while raising a heavy object away from the body's centre of gravity; or to accomplish pre-programmed, selective and accurate sequences of movement, as in playing a musical instrument, remain a source of wonder to the ordinary person and to the neuroscientist alike. How is this accomplished? Much can be learned by introspective, philosophical analysis, but, unfortunately, many of the problems thus formulated cannot yet be addressed satisfactorily. Diseases of the nervous system, when lesions are definable within the nervous pathways, offer the possibility of understanding the functional capabilities of the remaining parts. However, this concept can only be addressed when the remaining parts are themselves known both functionally and anatomically. This goal remains remote.

Hughlings Jackson tackled the description of disordered motor function phenomenologically, gradually building up a library of observations upon which he grafted analytical concepts derived from his feeling for an underlying philosophy of the organisation of things based on evolutionary concepts (see Chap. 1). He saw that the brain was a complex organ, capable of multiple functions, and recognised that in disease, when the brain was damaged, it was possible to discern fragments of functions that were, in the normally functioning brain, hidden from the observer by the interposition of higher levels of function. These higher levels were, in Jackson's view, laid upon the lower levels so that they subsumed them and yet were, in some measure, dependent on them (Taylor 1931/32). Jackson thus regarded brain development as an evolution of function, and brain damage as a dissolution of function. It was his great contribution to continue to regard these concepts critically throughout his life and to try to test them by clinical observation (Riese 1950). His particularly detailed studies of epilepsy were especially important in this regard in that he was able to observe the onset of focal motor seizures, and the spread or march of the excitatory event by the sequence of motor actions during the episode.

Jackson was rigorous in his interpretation of his observations in that he did not allow himself to extend his hypotheses beyond his data. Thus he treated his observations as relating to function and not to structure. This rigidity of approach led to complexity in his writing, but it allowed him to begin to explore the greater problem of the nature of psychological action in the brain. Again, it was his observations in epilepsy that provided insight into this, especially temporal lobe epilepsy, in which he noted the inappropriate recall of fragments of memory, and of visual, auditory and olfactory experience, and even the onset of emotional experiences during seizures. The similarity of these fragmented experiences to the fragments of motor actions that he observed during motor seizures did not escape him. Furthermore, he was able to compare these fragmented actions, whether motor or psychological, with the fragments of motor acts that remain available to the patient after a motor stroke, or with the fragments of speech utilisable by the patient with motor aphasia. He was much concerned with the problem of how large or complete were these fragments, whether they were always the same in different individuals, or in different pathological disorders of the brain, and thus whether they represented primordial fragments of evolutionary behaviour. Only rarely did he indicate that the anatomical location of these actions was a problem. Indeed, it is a well-known Jacksonian aphorism, derived from his observations in focal epilepsy, that movements, not muscles, are located in the cortex. This concept implies that movements involving different patterns of recruitment of individual muscles must be represented, whether inherently or in response to learning, in overlapping cortical circuits. In recent years these cortical representations of movements have been termed "motor programs". Indeed, a similar approach has been used in defining the cortical representation of somatic, visual and auditory sensation (Mountcastle et al. 1975), and this has been found to be closely linked to the motor system.

Since Jackson's time a number of concepts have come into regular use in attempts to address the problem of the nature of the processes that initiate and control movement, and these merit further consideration and definition. Neurologists are particularly interested in achieving understanding of the abnormalities observed in patients with lesions of the nervous system. This remains a viable approach to the wider problem of understanding motor control, since, even at its simplest conceptual level, lesions in man represent an opportunity to study the effects on motor control of damage to various parts of the nervous system (Merton 1981).

Effects of Lesions in Man

The conventional clinical classification of disorders of the motor system stresses the emphasis placed on lesions of the classic motor pathways, particularly the lower and upper motor neuron components.

Lower Motor Neuron Lesions

Lesions of the lower motor neuron seem relatively easy to understand in physiological terms, with resultant effects of weakness, absence of stretch reflexes, muscular atrophy, reduced tone and fasciculation, although the precise mechanisms governing atrophy remain a considerable problem. However, the disturbances of functional organisation that may result from selective damage to neurons within functional pools of motor neurons in the anterior horn have so far received little attention (see p. 120).

Upper Motor Neuron Lesions

Lesions of the upper motor neuron introduce a new level of complexity. Although there is a weakness, this is found in a particular distribution, especially involving distal muscles and proximal flexors, relatively sparing antigravity muscles. Further, there is increased muscular tone, reflecting release of monosynaptic and long-loop stretch reflexes from suprasegmental modulation, causing both static and dynamic effects that result in the clinical phenomena of spasticity: this affects either antigravity or flexor muscles in proportion to the activity of other descending connections with the otolith and labyrinthine systems. The tendon stretch reflexes are increased in their responsiveness. Thus, the upper motor neuron lesion displays the combination of positive, released and negative symptoms described in such detail by Hughlings Jackson. Attempts to establish the anatomical basis of these features have foundered in the complexity of the variability of the responsible lesions. Indeed, attempts to verify the pyramidal location of the lesion responsible for the upper motor neuron syndrome have themselves produced surprising results in that the location of the lesion is not necessarily limited to the corticospinal tracts but almost invariably involves adjacent non-pyramidal components of the descending motor system (Liddell and Phillips 1944; Lawrence and Kuypers 1968a, b).

Parkinson's Disease

Lesions at higher anatomical levels in the central nervous system are associated with more complex disorders of motor function. In Parkinson's disease there is a combination of negative and positive symptoms and signs, consisting respectively of akinesia and loss of postural responses, and of tremor and rigidity. Investigation of akinesia has proved particularly rich in deriving concepts of the underlying functional organisation of the motor system (Marsden 1982). The patient with extrapyramidal akinesia experiences great difficulty in initiating movements, described by Parkinson as "a peculiar disinclination to move". A similar absence of spontaneous movement is sometimes noted in patients with disorders of the prefrontal regions of the brain, but the latter differs in that movement is possible in relation to certain types or categories of task, although apparently absent in relation to others, thus indicating its category-specific relation, a feature characteristic of apraxia (Denny-Brown and Yanagisawa 1976). In Parkinsonian akinesia all voluntary movements are inhibited, inaccessible to command by an observer or in response to the willed effort to move. Yet the ability to move in certain situations is normal, as shown by the response to visual cues in walking, by the ability to get up and run in response to an urgent situation, or by the ability to carry out complex sequential motor tasks as in playing with a football, despite the apparently fixed dystonic posture of the disease. These clinical observations clearly imply that the motor programs of the brain are intact, but inaccessible or available only in part for everyday activities. The nigro-striatal pathway is believed to be crucial for activating these programs, and so in allowing normal movement sequences to be utilised (Marsden 1982). It is important to recognise that during activation of these sequences of motor programs in movement the Parkinsonian patient loses the dystonia and rigidity that is ordinarily so characteristic of the disease. Further evidence that the disease is primarily a disturbance of the ability to access and run motor programs comes from clinical observations of the effects of treatment with levodopa. During treatment the patient's motor abilities, including both postural reactions and the ability to fulfill voluntary motor tasks, are strikingly improved for the effective period of the drug's availability to the brain. The dopaminergic pathway, presumably the nigro-striatal component of this pathway, is thus important in switching between and within motor programs.

Apraxia

The motor programs themselves are intact in Parkinson's disease; they must therefore be stored in parts of the brain that are not damaged in the disease. Does the clinical concept of apraxia help in understanding this problem? Jackson regarded the highest level of nervous function as the premotor cortex, the middle level being the precentral and postcentral sensorimotor cortical areas. In ideomotor apraxia, as described by Liepmann (1900), there is degradation of the ability to formulate the different steps required to execute a movement, so that certain aspects of the movement may be performed adequately, but others are imperfect. Such patients are sometimes able to copy a meaningful movement sequence, or to formulate and execute it if given an appropriate context, and particularly if given the instruments usually used in that task. Thus the patient might be unable to pretend to strike a match but could use a matchbox normally. In ideational apraxia the holistic concept of the movement is itself lost. These disorders are associated with disease in the prefrontal cortex, but a similar motor dysfunction, called by Hecaen "apractagnosia" (Hecaen et al. 1956), is recognised in patients with right parietal disease, as part of the spatial and conceptual disorder found in that syndrome.

This simple model of a highest level, consisting of motor and sensory programs embedded in a cortical system that is itself subservient in some way to the strategic activities of the prefrontal brain, communicating to effector organs in the lowest centre, the brain stem and spinal cord, organised in terms of muscles rather than movements, is strictly Jacksonian, and clearly an oversimplification. Indeed, it implies that the higher centres, or levels, are always in a command relationship to the lower. Again, clinical observation illustrates other aspects of these systems.

Involuntary Movement Disorders

The involuntary movement disorders indicate that there are multiple levels at which communication between emerging motor programs may occur, and also that communication between sensory and motor systems is not limited to cerebral cortex, but must occur in other parts of the nervous system. In patients with chorea, an involuntary movement disorder characterised by rapid ballistic movements interposed in ordinary motor tasks, it is striking that despite the rapidity and amplitude of the involuntary movements the patient's own voluntary movements may be little disturbed in their accuracy. In this disorder there is no bradykinesia, implying that motor programs are accessible for recompilation, assembly and execution, and the normality of their execution suggests that there is sufficient redundancy in their structure to enable random perturbations caused by the interposed involuntary movement to be compensated, resulting in an accurate movement.

In more severe disorders, however, as in hemichorea, the involuntary movement is too great to enable the central nervous system to compensate, and motor performance is degraded. Redundancy implies that alternative strategies are available for course and velocity correction, and that the movement is monitored during its execution, or that the involuntary movement itself is biphasic so that it does not result in any abnormality of the intended movement. Since the time relationships of the intended movement must be distorted by the interposed involuntary movement, the latter cannot be the case.

Error Monitoring and Correction

The commonly experienced problem of interruption of motor programs during their performance is relevant to the question of internal monitoring of motor performance. If a rapidly performed, learned movement sequence, such as playing the piano, is attempted after a few days' rest it is common for the task to be interrupted by an error, detected not only by the incorrect sound sequence itself, but by an internally perceived feeling of a mistake. The subsequent sequence of movements may be so disturbed that the task may have to be interrupted before it can be resumed. This is the correlate of the concept of efference copy, by which the sensorimotor cortex is considered to monitor the output of the motor cortex by matching the output against the internally held program for the sequence of movements. This concept is particularly useful in clinical neurology because it is consistent with observations of the importance of sensory mechanisms in planning and executing voluntary movements, as in those dependent on parietal cortex. Profound disturbances of voluntary movement also occur with lesions in the posterior columns (Rothwell et al. 1982), an observation illustrating the interaction of sensory and motor activity in infracortical parts of the central nervous system.

Organisation of Motor Programs in the Brain

Are there general rules governing the sequencing of motor performance? Programs subserving the earliest acquired motor programs tend to be organised at a more peripheral level. Thus chewing is a function of rhythm generators and of sensorimotor integration in the brain stem; stepping can be generated in the spinal cord of the newborn infant: and even in the presence of anencephaly certain forelimb movements can be initiated by activity in motor neurons in the cervical segments. Spinal neurons receive complex inputs from propriospinal and cutaneous afferents in addition to input from descending pathways signalling information from higher levels in the neuraxis. Thus, there is evidence of a distributive system of control of motor performance, functioning at multiple anatomical levels in the nervous system. Only by these means can the lowest common units of motor performance, the anterior horn cells themselves, be coordinated into complex and different sequences of activities suitable for different tasks. Individual muscles can be used for different tasks, including activity as agonist or antagonist, phasic or tonic movements, or in static or dynamic tasks (Phillips and Porter 1977).

Posture and Movement

Postural adjustment always precedes voluntary movement. Unless this were the case, falls would be common. Indeed sudden falls in the elderly may, in part, result from disturbances in this relationship caused by faults in the systems that ordinarily ensure this relationship. Posture is maintained by the coactivation and integration of reflexes that are mainly dependent on spinal, brain stem and basal ganglia function. These reflexes feed downstream and upstream to sensorimotor maps in tectal brain stem, cerebellum, spinal cord, basal ganglia and cerebral cortex, where modulation of the activity of postural reflexes is carried out in relation to the anticipated movement of the limbs or trunk. The anticipatory processing necessary for planning the movement sequence itself (Tanji and Evarts 1976) seems to take place during a period perhaps as long as a second in the supplementary motor area (Deecke 1987), in association with motor areas of the cortex and the basal ganglia (Marsden 1987). The tectal map in the superior colliculi is especially interesting in this regard. Here a sensory visual map of the retina is connected to a second map related to goal points of saccadic eye movements, which are unrelated to the initial eye position (Buttner-Ennever and Buttner 1978; Sparks and Mays 1983). Thus this arrangement can be used to calculate vectors for horizontal or vertical saccades that are themselves generated in the parapontine reticular formation or in the midbrain reticular formation, respectively. This mechanism has clear potential applications for the calculation of vectors for kinetic movements of the upper limbs in ballistic and reaching tasks that are carried out in the brain and spinal cord. Postural set for such movements is controlled elsewhere in the brain stem and basal ganglia, presumably from a library of previously acquired programs (Martin 1967).

Program Ordering

In considering the ordering of motor programs it is thus evident that there are a number of steps that are necessary in their organisation. There must be a program memory; this involves both sensory and motor components. Subroutines of motor programs must be available for access and for assembly into larger programs for everyday use. Certain sets of programs will be used in conjunction, and these will require alternative forms for similar but slightly different tasks. This redundancy will reduce the likelihood of error in execution of the task. Indeed, lack of redundancy, for example in the damaged nervous system or during the process of learning a new motor task, has been suggested as a mechanism for the generation of the feeling of fatigue so characteristic of these situations.

The execution of the program is initiated by the brain, perhaps by one of several possible levels of nervous function rather than solely from the prefrontal cortex, and is monitored by the brain. Monitoring by efference copy and by visual, auditory, labyrinthine and proprioceptive input are all relevant to different degrees in different tasks. However, monitoring performance by input from the periphery is likely to introduce an unacceptable delay into the process with negative feedback that will lead to cumulative errors in accuracy. These factors may therefore be more relevant in monitoring learning, and in recomputing internal steps in the motor program for a complex task, rather than in everyday execution or performance of the program.

Role of the Cerebellum

How does the cerebellum contribute to motor performance? Current concepts of the cerebellum illustrate rather well the dichotomy between clinical observations of lesions in a system and understanding of its function in the normal individual. The phenomenology of cerebellar disease is relatively well documented. Holmes (1939) suggested that the cerebellum might function in error detection, since it receives input from all sensory pathways, together with extensive connections to the motor systems involved in axial, proximal and distal limb muscles, and to the vestibulo-labyrinthine system. It has been proposed that the cerebellum is particularly concerned with the acquisition of motor skills (Marr 1969). This suggestion is inherently unsatisfactory for the neurologist confronted with patients unable to stand from ataxia due to cerebellar lesions, and a role of this organ in the execution of motor activity is self-evident. Abnormalities in the modulation of the vectors utilised in voluntary and in postural movements are evident in patients with cerebellar disease, and these suggest that the cerebellum plays a role in the compilation of the motor program. The smooth performance of the program is disturbed by inaccuracies in these vectors that cannot be modified during their execution, and that cannot subsequently be corrected when another attempt is made to carry out the movement. In watching a patient with cerebellar ataxia trying to perform a smoothly directed reaching task, or a rapidly alternating movement, it is often striking that the patient is surprised that the movement is so badly carried out. This observation raises the possibility that the sensorimotor program is perceived by the brain as accurate, but that its transposition to the effector parts of the motor system and its execution are disturbed by the cerebellar lesion.

Task-groups of Neurons

It is known from the work of Sherrington, and of Sharrard (1955), that motor cells in the anterior horn of the spinal cord are arranged in functional columns that represent muscles (Swash et al. 1986). This is in accord with Jackson's notion that movements are represented in the highest level, and muscles in the lowest level. Stuart et al. (1988) have pointed out that the functional significance of the architectural arrangement of motor cells in the cord may be of fundamental importance. Segmental motor control mechanisms may be important in simplifying the task of the central nervous system in formulating strategies for motor control. This concept has been proposed in terms of task-groups of neurons in the spinal cord, including their related sensory neurons and receptors. These task-groups may be multiple within neurons forming part of the nuclear grouping of individual muscles, so that functional rearrangements within the anterior horn enable different types of kinetic tasks to be achieved. Thus the centrally generated motor programs may be oriented to direct activity in pre-programmed task-groups of motor and sensory neurons at spinal level. This notion further expands the possibilities for distributive relationships of the hierarchies of the motor control system, reaffirming the concept of multiple sites for direction of components of motor program execution in motor control.

Conclusion

Thus it is not likely that there is an overall Sherringtonian "head ganglion" of the nervous system in motor control, but rather that motor control is achieved by interactions at multiple anatomical and physiological levels, using a combination of pre-programmed task-groups of sensory and motor neurons in fluctuating arrays of functional units. There is much redundancy in the system in health, enabling a given task to be performed by one or more of several different strategies (Ito 1986). Ageing and disease may reduce the available options for effective motor program assembly, by introducing uncorrectable errors into the programs, or into their assembly, transposition through the nervous system, or execution, leading to the clinical phenomena of disordered motor control. Such lesions may also cause disruption of feedback loops used for error monitoring and correction during the smooth coactivation of the distributive systems that are utilised in this complex learned and inbuilt motor control system. Thus the Jacksonian hierarchies are currently conceived as control systems that extend through multiple levels of the nervous system, and the presence of particular clinical anomalies resulting from certain lesions does not necessarily imply that the structures injured are directly concerned with the abolition or suppression, or with the release, of the particular functions that appear deranged by the lesion. Indeed, similar clinical anomalies may result from lesions at different anatomical levels in the central nervous system, and they may themselves be dependent on the emotional state, and on other aspects of the general state of the patient at the time of the examination. Thus the negative and positive features so clearly elucidated by Jackson can be closely related to modern neurophysiological and neurological observations.

References

Buttner-Ennever JA, Buttner U (1978) A cell group associated with vertical eye movement in the rostral mesencephalic reticular formation of the monkey. Brain Res 151:31–47

Deecke L (1987) Bereitschaftspotential as an indicator of movement preparation in supplementary

motor area and motor cortex. In: Motor areas of the motor cortex. Ciba Foundation Symposium 132. Pitman, London, pp 251–264.
Denny-Brown D, Yanagisawa N (1976) The role of the basal ganglia in the initiation of movement. In: Yahr MD (ed) The basal ganglia. Raven, New York, pp 115–148
Ferrier D (1876) The functions of the brain. Smith, Elder, London, p 323
Hecaen H, Penfield W, Bertrand C, Malmo R (1956) The syndrome of apractagnosia due to lesions of the minor cerebral hemisphere. Arch Neurol Psychiatry 75:400–434
Holmes G (1939) The cerebellum of man. Brain 62:1–30
Ito M (1986) Neural control systems controlling movement. Trends Neurosci 100:515–518
Lawrence DG, Kuypers HGJM (1968a) The functional organization of the motor system in the monkey: 1. The effects of bilateral pyramidal lesions. Brain 91:1–14
Lawrence DG, Kuypers HGJM (1968b) The functional organization of the motor system in the monkey: 2. The effects of lesions on the descending brain-stem pathways. Brain 91:15–36
Liddell EGL, Phillips CG (1944) Pyramidal section in the cat. Brain 67:1–9
Liepmann H (1900) Das Krankheitshild der Praxie (motorenischen Asymbolie). Monatsschr Psychiatrie 8:15–44, 102–132, 182–197
Marr D (1969) A theory of cerebellar cortex. J Physiol (Lond) 202:437–470
Marsden CD (1982) The mysterious motor function of the basal ganglia. Neurology 32:514–540
Marsden CD (1987) What do the basal ganglia tell premotor cortical areas? In: Motor areas of the cerebral cortex. Ciba Foundation Symposium 132. Pitman, London, pp 296–300
Martin JP (1967) The basal ganglia and posture. Pitman Medical, London
Merton PA (1981) Neurophysiology in man. J Neurol Neurosurg Psychiatry 44:861–871
Mountcastle VB, Lynch JC, Georgopoulos A, Sakata H, Acuna C (1975) Posterior parietal association cortex of the monkey: command functions for operations within extrapersonal space. J Neurophysiol 38:871–908
Phillips CG, Porter R (1977) Corticospinal neurones. Academic, London
Riese W (1950) Principles of neurology, in the light of history and their present use. Nervous and Mental Disease Monographs, New York, p 177
Rothwell JC, Traub MM, Day BL, Obeso JA, Thomas PK, Marsden CD (1982) Manual performance in a deafferented man. Brain 105:515–542
Sharrard WJW (1955) The distribution of the permanent paralysis in the lower limb in poliomyelitis; a clinical and pathological study. J Bone Joint Surg [Br] 37:540–558
Sparks DL, Mays LE (1983) Spatial localisation of saccade targets. 1. Compensation for stimulation-induced perturbations in eye position. J Neurophysiol 49:45–63
Stein JF (1986) The control of movement. In: Coen C (ed) Functions of the brain. Clarendon, Oxford, pp 67–97
Stuart DG, Hamm TM, Noven SV (1988) Partitioning of monosynaptic Ia EPSP connections with motoneurons according to neuromuscular topography; generality and functional implications. Prog Neurobiol 30:437–447
Swash M, Leader M, Brown A, Swettenham KW (1986) Focal loss of anterior horn cells in the cervical cord in motor neuron disease. Brain 109:939–952
Tanji J, Evarts EV (1976) Anticipatory activity of motor cortex neurons in relation to direction of an intended movement. J Neurophysiol 39:1062–1068
Taylor J (ed) (1931/32) Selected writings of John Hughlings Jackson, vols 1 and 2. Hodder and Stoughton, London. Reprinted (1958) Basic Books, New York

Chapter 15

Information Processing and Basal Ganglia Function

E. T. Rolls

Some parts of the striatum receive inputs from parts of the brain at the highest level of information processing such as the association cortex and limbic structures. The first question considered in this chapter is whether the basal ganglia provide a route for information from these highest levels of information processing to influence motor function, given that the striatum has outputs through other parts of the basal ganglia to structures concerned with motor function such as the premotor cortex and the supplementary motor area. Some parts of the striatum also receive inputs from structures concerned with movement such as the motor cortex and the somatosensory cortex. A second question considered, therefore, is whether there is any mixing or integration within the striatum of the inputs it receives from the parts of the cortex concerned with high-level information processing and from the limbic system with inputs received from the somatosensory and motor cortices. These questions are considered using evidence from both the anatomy and the neurophysiology of the basal ganglia. Before leaving the topic of the hierarchical organisation of information processing in each of the sensory systems, and of what is achieved by the hierarchical organisation found, I wish to note that these questions, and the type of coding found in such neural systems, are discussed elsewhere (Rolls 1987a; 1988).

Connections of the Basal Ganglia

The point-to-point connectivity of the basal ganglia, as shown by experimental anterograde and retrograde neuroanatomical path tracing techniques in the primate, provides some evidence for at least partial segregation of pathways within the basal ganglia. The segregation can be described as forming three divisions (Figs. 15.1, 15.2) (see Kemp and Powell 1970, 1971; Heimer and Wilson 1975; Tobias 1975; Van Hoesen et al. 1976; Newman and Winans 1980a,b; Hemphill et al. 1981; Heimer et al. 1982; Carpenter 1984; DeLong et al. 1984;

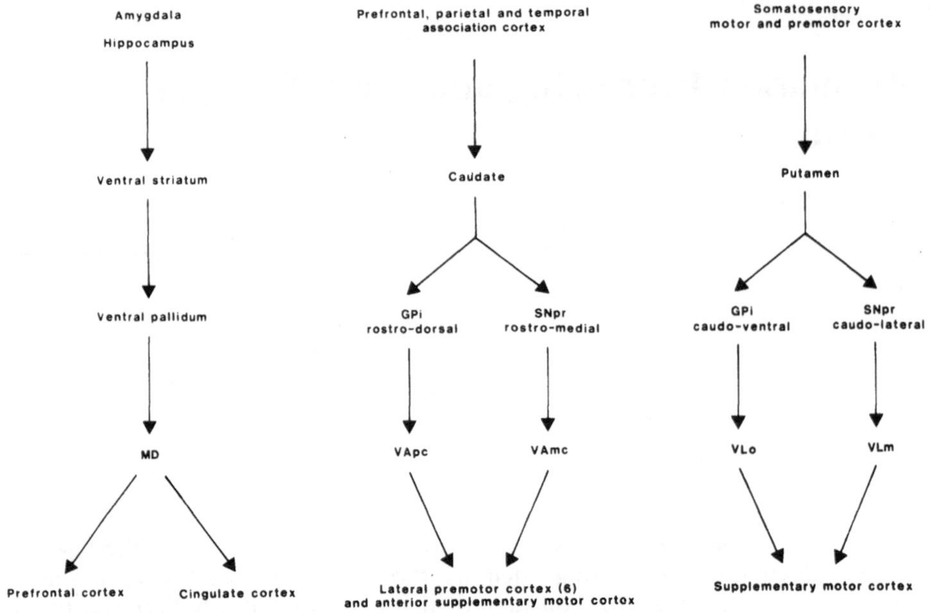

Fig. 15.1. A synthesis of some of the anatomical studies (see text) of the connections of the basal ganglia. *GPi*, globus pallidus, internal segment; *MD*, nucleus medialis dorsalis; *SNpr*, substantia nigra pars reticulata; *VAmc*, n. ventralis anterior pars magnocellularis; *VApc*, n. ventralis anterior pars compacta; *VLo*, n. ventralis lateralis pars oralis; *VLm*, n. ventralis pars medialis.

Jurgens 1984; Nauta and Domesick 1984; Schell and Strick 1984; Seleman and Goldman-Rakic 1985; Webster 1988). First, the motor cortex (area 4) and somatosensory cortex (areas 3, 1 and 2) project somatotopically to the putamen, which has connections through the globus pallidus and substantia nigra to the ventral anterior thalamic nuclei and thus to the supplementary motor cortex (see Fig. 15.1). Second, the prefrontal, parietal and temporal association cortices project to the caudate nucleus, which has connections through parts of the globus pallidus and substantia nigra to the ventral anterior group of thalamic nuclei and thus to the lateral premotor cortex (area 6). Third, limbic structures such as the amygdala and hippocampus project to the ventral striatum, which has connections through the ventral pallidum to the mediodorsal nucleus of the thalamus and thus to the prefrontal and cingulate cortices. Thus anatomical studies provide evidence for segregation of at least three major systems through the basal ganglia, with each system finally connecting to different cortical areas. There is even evidence for some segregation within each of these systems, at least in the striatum. For example, studies in the monkey show that the motor cortex (area 4) and somatosensory cortex (areas 3, 1 and 2) project somatotopically to the putamen (Kemp and Powell 1970, 1971; Kunzle 1975, 1977, 1978; Kunzle and Akert 1977; Jones et al. 1977), each region of the association cortex projects to a longitudinal strip of the caudate nucleus, with a dorsolateral strip of the caudate receiving from the posterior parietal cortex, a central strip receiving from the dorsolateral prefrontal cortex, and a ventromedial strip receiving from the

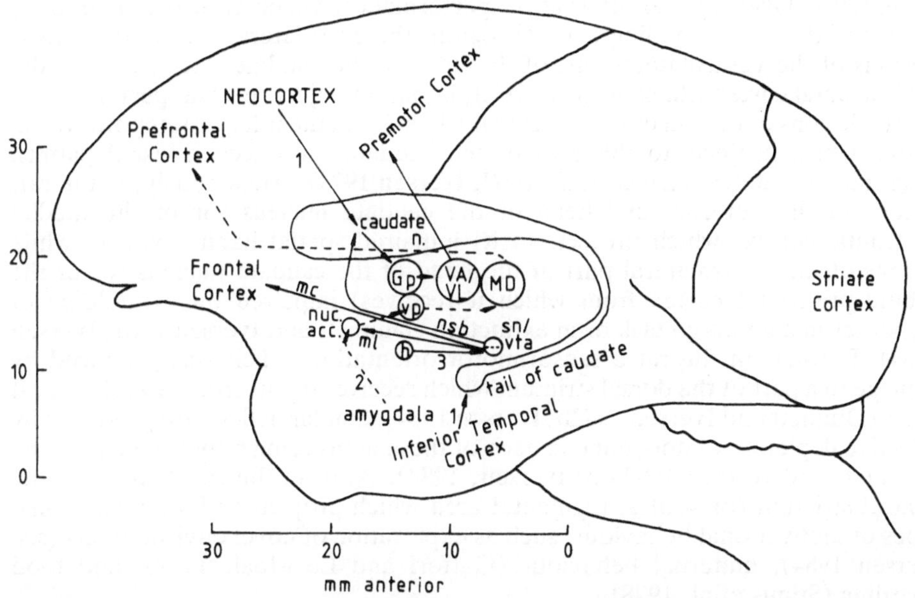

Fig. 15.2. Some of the striatal and connected regions in which the activity of single neurons is described, shown on a lateral view of the brain of the macaque monkey brain. *Gp*, globus pallidus; *h*, hypothalamus; *sn*, substantia nigra pars compacta (A9 cell group), which gives rise to the nigro-striatal dopaminergic pathway, or nigro-striatal bundle (*nsb*); *vta*, ventral tegmental area, containing the A10 cell group, which gives rise to the mesocortical dopamine pathway (*mc*) projecting to the frontal and cingulate cortices and to the mesolimbic dopamine pathway (*ml*), which projects to the nucleus accumbens (*nuc acc*). *MD*, nucleus medialis dorsalis; *VA/VL*, nucleus ventralis and lateralis of the thalamus; *VP*, ventral pallidum.

orbitofrontal, anterior cingulate and superior temporal cortices (Seleman and Goldman-Rakic 1985).

Effects of Striatal Lesions

Damage to the striatum produces effects which suggest that it is involved in orientation to stimuli and in the initiation and control of movement. In animals, lesions of the dopamine pathways which deplete the striatum of dopamine lead to a failure to orient to stimuli, a failure to initiate movements which is associated with catalepsy, and a failure to eat and drink (Marshall et al. 1974). In man, depletion of dopamine in the striatum is found in Parkinson's disease, in which there is tremor and akinesia, that is, a lack of voluntary movement (Hornykiewicz 1973). However, consistent with the anatomical evidence, the effects of damage to different regions of the striatum also suggest that there is functional specialization within the striatum (Divac and Oberg 1979; Oberg and Divac 1979; Iversen 1984). The selective effects may be related to the function of the cortex or limbic structure from which a region of the striatum receives inputs. For example, in the monkey, lesions of the anterodorsal part of the head of the caudate nucleus disrupted delayed spatial alternation performance, which is also impaired by lesions of the corresponding cortical region, the dorsolateral prefrontal cortex. Lesions of the ventrolateral part of the head of the caudate nucleus (as of the orbitofrontal cortex which projects to it), impaired object reversal performance. Lastly, lesions of the tail of the caudate nucleus (as of the inferior temporal visual cortex which projects to this part of the caudate) produced a visual pattern discrimination deficit (Divac et al. 1967; Iversen 1979). Analogously, in the rat, lesions of the anteromedial head of the caudate nucleus (or of the medial prefrontal cortex, which projects to it) impaired spatial habit reversal, while lesions of the ventrolateral part of the head of the caudate nucleus (or of the orbital prefrontal cortex from which it receives) impaired the witholding of responses in a go/no-go task or in extinction (Dunnett and Iversen 1981; Iversen 1984). Further, in the rat a sensorimotor orientation deficit was produced by damage to a part of the dorsal striatum which receives inputs from lateral cortical areas (Dunnett and Iversen 1982b; Iversen 1984). Similar deficits are produced by selective depletion of dopamine in each of these areas using 6-hydroxydopamine (Dunnett and Iversen 1982a,b; Iversen, 1984). Also, in the rat damage to the ventral striatum (or ventral tegmental area which projects to it) impairs some types of motivational behaviour, such as exploration of novel environments (see Iversen 1984), maternal behaviour (Gaffori and Le Moal, 1979), and food hoarding (Stinus et al. 1978).

Neuronal Activity in the Striatum

The evidence from the connections of the striatum and from the effects of damage to the striatum thus suggests that there may be functional segregation within the striatum. To investigate this more directly, the activity of single neurons has been recorded in different parts of the striatum of behaving monkeys in tasks known to be impaired by damage to the striatum. These investigations are described next. These tasks included the initiation of behaviour, for example the initiation of feeding and of limb movements; and visual discrimination performance and its reversal (see Rolls 1984b).

Head of the Caudate Nucleus

The activity of 394 neurons in the head of the caudate nucleus and most anterior part of the putamen was analysed in three behaving rhesus monkeys (Rolls et al. 1983). Of these neurons 64.2% had responses related to environmental stimuli, movements, the performance of a visual discrimination task, or eating. However, only relatively small proportions of these neurons had responses which were unconditionally related to visual (9.6%), auditory (3.5%), or gustatory (0.5%) stimuli, or to movements (4.1%). Instead, the majority of the neurons had responses which occurred conditionally in relation to stimuli or movements, in that the responses occurred in only some test situations, and were often dependent on the performance of a task by the monkeys. Thus, it was found that in a visual discrimination task 14.5% of the neurons responded during a 0.5 s tone/light cue which signalled the start of each trial; 31.1% responded in the period in which the discriminative visual stimuli were shown, with 24.3% of these responding more to either the visual stimulus which signified food reward or to that which signified punishment; and 6.2% responded in relation to lick responses. Yet these neurons typically did not respond in relation to the cue stimuli, to the visual stimuli, or to movements, when these occurred independently of the task or performance of the task was prevented. Similarly, although of the neurons tested during feeding, 25.8% responded when the food was seen by the monkey, 6.2% when he tasted it, and 22.4% during a cue given by the experimenter that a food or non-food object was about to be presented, only few of these neurons had responses to the same stimuli presented in different situations. Further evidence on the nature of these neuronal responses was that many of the neurons with cue-related responses only responded to the tone/light cue stimuli when they were cues for the performance of the task or the presentation of food, and some responded to the different cues used in these two situations.

The finding that such neurons may respond to environmental stimuli only when they are significant (Rolls et al. 1983) has now been confirmed by Evarts and his colleagues. They showed that some neurons in the putamen only responded to the click of a solenoid when it indicated that a fruit juice reward could be obtained (see Evarts and Wise 1984). We have found that this decoding of the significance of environmental events which are signals for the preparation for or initiation of a response is represented in the firing of a population of neurons in the dorsolateral prefrontal cortex, which projects into the head of the caudate nucleus (E. T. Rolls and G. C. Baylis 1984, unpublished observations). These neurons respond to the tone cue only if it signals the start of a trial of the visual discrimination task, just as do the corresponding population of neurons in the head of the caudate nucleus. The finding that the decoding of significance is performed by the cortex, and that the striatum receives only the result of the cortical computation, is considered below and elsewhere (Rolls and Williams 1987).

These findings indicate that the head of the caudate nucleus and most anterior part of the putamen contain populations of neurons which respond to cues which enable preparation for the performance of tasks such as feeding and tasks in which movements must be initiated, and others which respond during the performance of such tasks in relation to the stimuli used and the responses made, yet that the majority of these neurons do not have unconditional sensory or motor responses. It has therefore been suggested (Rolls et al. 1983) that the anterior neostriatum contains neurons which are important for the utilisation of environmental cues for the preparation for behavioural responses, and for particular behavioural responses made in particular situations to particular environmental stimuli, that is, in stimulus–motor response habit formation. Different neurons in the cue-related group often respond to different subsets of environmentally significant events, and thus convey some information which would be useful in switching behaviour and in preparing to make different responses.

It may be suggested that deficits in the initiation of movements following damage to striatal pathways may arise in part because of interference with these functions of the anterior neostriatum. Thus the akinesia or lack of voluntary movement produced by damage to the dopaminergic nigro-striatal bundle in animals and present in Parkinson's disease in man (Hornykiewicz 1973) may arise at least in part because of dysfunction of a system which normally is involved in utilising environmental stimuli which are used as cues in the preparation for the initiation of movements. Such preparation may include, for example, postural adjustments. The movement disorder may also be due in part to the dysfunction of the system of neurons in the head of the caudate nucleus which appears to be involved in the generation of particular responses to particular environmental events (see also Rolls et al. 1979; Rolls et al. 1983).

Tail of the Caudate Nucleus

The projections from the inferior temporal cortex and the prestriate cortex to the striatum arrive mainly, although not exclusively, in the tail of the caudate nucleus and in the posteroventral portions of the putamen (Whitlock and Nauta 1956; Kemp and Powell 1970; Yeterian and Van Hoesen 1978; Van Hoesen et al. 1981). Since these regions of the caudate nucleus and putamen are adjacent and have a common anatomical input they are referred to together as the caudal neostriatum. Although there is this visual projection directly into the caudal neostriatum, there have been few studies of the functions of these visual pathways and their importance in visually controlled behaviour. Divac et al. (1967) reported that stereotaxic lesions placed in the tail of caudate nucleus in the region which receives input from the inferior temporal visual cortex produced a deficit in visual discrimination learning. The lesion did not produce impairment in an object reversal task, though in two out of four monkeys it did disturb delayed alternation performance. Buerger et al. (1974) found that lesions of the ventral putamen in the monkey produced a deficit in the retention of a preoperatively learnt visual discrimination problem, but the lesion did not disturb retention of auditory discrimination or delayed alternation tasks. The deficit produced by both neostriatal lesions seems therefore to reflect predominantly a loss of visual functions rather than a general loss of cognitive functions.

Since so little is known about the nature of visual processing within the caudal neostriatum and the fate of the visual input from inferior temporal cortex to this area, the activity of single neurons was recorded in the tail of the caudate nucleus and adjoining part of the ventral putamen (Caan et al. 1984). Of 195 neurons analysed in two macaque monkeys, 109 (56%) responded to visual stimuli, with latencies of 90–150 ms for the majority of the neurons. The neurons responded to a limited range of complex visual stimuli, and in some cases responded to simpler stimuli such as bars and edges. Typically (in 75% of cases) the neurons habituated rapidly, within 1–8 exposures, to each visual stimulus, but remained responsive to other visual stimuli with a different pattern. This habituation was orientation specific, in that the neurons responded to the same pattern shown in an orthogonal orientation. The habituation was also relatively short-term, in that at least partial dishabituation to one stimulus could be produced by a single intervening presentation of a different visual stimulus. These neurons were relatively unresponsive in a visual discrimination task, having habituated to the stimuli which had been presented in the task on many previous trials.

The main characteristics of the responses of these neurons in the tail of the caudate nucleus and adjoining part of the putamen were rapid habituation to specific visual patterns, their sensitivity to changes in visual pattern, and the relatively short-term nature of their habituation to a particular pattern, with dishabituation occurring to a stimulus by even one intervening trial with another stimulus. Given these responses, it may be suggested that these neurons are involved in short-term pattern-specific habituation to visual stimuli. This system would be distinguishable from other habituation systems (involved, for example, in habituation to spots of light) in that it is specialised for patterned visual stimuli which have been highly processed through visual cortical analysis mechanisms, as shown not only by the nature of the neuronal responses, but also by the fact that this system receives inputs from the inferior temporal visual cortex. It may also be suggested that this sensitivity to visual pattern change may have a role in alerting the monkey's attention to new stimuli. This suggestion is consistent with the changes in attention and orientation to stimuli produced by damage to the striatum. Thus, damage to the dopaminergic nigro-striatal bundle produces an inability to orient to visual and other stimuli in the rat (Marshall et al. 1974).

Putamen

The putamen receives inputs from the sensorimotor cortex, areas 3,2,1,4 and 6 (Kunzle 1975, 1977, 1978; Kunzle and Akert 1977; Jones et al. 1977; DeLong et al. 1983). It is clear that the activity of many neurons in the putamen is related to movements (Anderson 1978; Crutcher and DeLong 1984a,b; DeLong et al. 1984). There is a somatotopic organisation of neurons in the putamen, with separate areas containing neurons responding to arm, leg or orofacial movements. Some of these neurons respond only to active movements, and others to active and to passive movements. Some of these neurons respond to somatosensory stimulation, with multiple clusters of neurons responding, for example, to the movement of each joint (see Crutcher and DeLong 1984a; DeLong et al. 1984). In experiments in which the arm has been given assisting and opposing loads, some neurons in the putamen have been shown to respond in relation to the direction of an intended movement, rather than in relation to the muscle forces required to execute the movement (Crutcher and DeLong 1984b). Also, the firing rate of neurons in the putamen tends to be linearly related to the amplitude of movements (Crutcher and DeLong 1984b), and this is of potential clinical relevance, since patients with basal ganglia disease frequently have difficulty in controlling the amplitude of their limb movements.

In order to obtain further evidence on specialisation of function within the striatum, the activity of neurons in the putamen has been compared with the activity of neurons recorded in different parts of the striatum in the same tasks (Rolls et al. 1984). Of 234 neurons recorded in the putamen of two macaque monkeys during the performance of a visual discrimination task and the other tests in which other striatal neurons have been shown to respond (Rolls et al. 1983; Caan et al. 1984), 68 (29%) had activity which was phasically related to movements (Rolls et al. 1984). Many of these responded in relation to mouth movements such as licking. The neurons did not have activity related to taste, in that they responded, for example, during tongue protrusion made to a food or non-food objects. Some of these neurons responded in relation to the licking mouth movements made in the visual discrimination task, and always also responded when mouth movements were made during clinical testing when a food or non-food object was brought close to the mouth. Their responses were thus unconditionally related to movements, in that they responded in whichever testing situation was used, and were therefore different from the responses of neurons in the head of the caudate nucleus (Rolls et al. 1983).

Of the 68 neurons in the putamen with movement-related activity in these tests, 61 had activity related to mouth movements, and 7 had activity related to movements of the body. Of the remaining neurons, 24 (10%) had activity which was task related, in that some change of firing rate associated with the presentation of the tone cue or the opening of the shutter occurred on each trial (see Rolls et al. 1984), 4 had auditory responses, 1 responded to environmental stimuli (see Rolls et al. 1983), and 137 were not responsive in these test situations.

These findings (Rolls et al. 1984) provide further evidence that differences between neuronal activity in different regions of the striatum are found even in the same testing situations, and also that the inputs which activate these neurons are derived functionally from the cortex which projects into a particular region of the striatum (in this case sensorimotor cortex, areas 3,1,2,4 and 6).

Ventral Striatum

To analyse the functions of the ventral striatum, the activity of more than 1000 single neurons has been recorded in a region which included the nucleus accumbens and olfactory tubercle in five macaque monkeys in test situations in which lesions of the amygdala, hippocampus and inferior temporal cortex produce deficits (Rolls et al. 1982; Rolls and Williams 1987). The following types of neuronal response have so far been identified (Table 15.1):

1. There are neurons which respond to novel visual stimuli. Different neurons in this category show pattern-specific habituation over 1–10 trials, and show retention of this habituation over 1–14 intervening trials. A smaller number of neurons responds to familiar rather than to novel visual stimuli. This first group of neurons thus has responses related to recognition memory, responding differently to novel and to familiar stimuli. They may receive their inputs from the amygdala and hippocampus, in both of which there are small numbers of neurons which respond differently to novel as compared with familiar visual stimuli (Rolls 1985, 1987a, b).

2. Other neurons respond to visual stimuli of emotional or motivational significance, that is to stimuli which have in common the property that they are positively or negatively reinforcing (Rolls 1986a). (Reinforcers are stimuli which, if their occurrence, termination or omission is made contingent upon the making of a behavioural response, alter the future emission of that response.) Of the neurons which responded to visual stimuli which were rewarding, relatively few responded to all the rewarding stimuli used. That is, only a few ventral striatal neurons responded both when food was shown and to the positive discriminative stimulus, S+, in a visual discrimination task, as shown in Table 15.1. Instead, the reward-related neuronal responses were typically more context- and stimulus-dependent, responding, for example, to the sight of food but not to the S+ which signified food ("food" in Table 15.1), differentially to the S+ or S− but not to food, or to food if shown in one context but not in another context. Some other neurons responded to aversive stimuli. These neurons do not respond simply in relation to arousal, produced by inputs from different modalities. These neurons

Table 15.1. Neuronal responses in the ventral striatum[a]

	Ventral striatum	
	No./1013	%
1. Visual, recognition related		
novel	39	3.5
familiar	11	1.1
2. Visual, association with reinforcement		
aversive	14	1.4
food	44	4.3
food and S+	18/1004	1.8
food, context dependent	13	1.3
opposite to food/aversive	11	1.1
differential to S+ or S− only	44/1112	4.0
3. Visual		
general interest	51	5.0
non-specific	78/1112	7.0
face	17	1.7
4. Movement-related, conditional	50/1112	4.5
5. Somatosensory	76/1112	6.8
6. Cue related	177/1112	15.9
7. Responses to all arousing stimuli	9/1112	0.8
8. Task-related (non-discriminating)	17/1112	1.5
9. During feeding	52/1112	4.7
10. Peripheral visual and auditory stimuli	72/538	13.4
11. Unresponsive	608/1112	54.7

[a] The sample size was 1013 neurons except where indicated. The categories are non-exclusive.

with reinforcement-related responses represented 13.9% of the neurons recorded in the ventral striatum, and may receive their inputs from structures such as the amygdala, in which some neurons with similar responses are found (Rolls 1985, 1986a; Sanghera et al. 1979). In addition, some neurons responded when the monkey looked at faces (Table 15.1). The neurons which respond to faces are probably part of a system of neurons important in the recognition of individuals by their faces and in social and emotional responses to faces. Neurons in this system are found in the cortex in the superior temporal sulcus of the macaque monkey, and in the amygdala, which receives from the temporal lobe cortex and projects to the nucleus accumbens (Rolls 1984a, 1988; Leonard et al. 1985; Baylis et al. 1985, 1987; Rolls and Williams 1987). Thus it is likely that the ventral striatum is one system which provides a route for this system to influence behaviour. It will thus be of interest to determine whether any of the social or emotional responses to faces and other stimuli known to be affected by damage to the amygdala (Rolls 1985, 1986a, 1988) are influenced when the functions of the ventral striatum are disrupted.

3. Other neurons with visual responses had activity which occurred primarily to objects which the monkey paid attention to in the environment ("Visual, general interest" in Table 15.1), or which occurred to all visual stimuli presented ("Visual, non-specfic" in Table 15.1). Altogether, the neurons with visual responses, defined according to the criteria of Sanghera et al. (1979), represented 32.2% of those recorded in the ventral striatum (categories 1–3 of Table 15.1), and probably reflected the inputs received from the inferior temporal cortex as well as from the amygdala and hippocampus by the ventral striatum (Rolls 1985).

4 and 5. Other neurons responded in relation to movement or somatosensory stimuli, for example during licking or arm movements. Some neurons appeared to have larger responses when the monkeys reached for food than when they reached towards other objects, and these are described as "Movement-related, conditional" in Table 15.1.

6. Other neurons responded, as in the head of the caudate nucleus (Rolls et al. 1983), to cues such as 0.5s tone which the monkey used to prepare for the performance of each trial of the visual discrimination task.

7. Other neurons responded in relation to arousal, however this was produced.

8. Some neurons responded while the monkey performed the visual discrimination task, but did not have differential responses to the rewarding and aversive stimuli in the task, and could not be shown to have visual or movement-related movements outside the task. These neurons are described as "task-related" in Table 15.1.

9. Some neurons responded when the experimenter fed the monkey either just before or during the eating.

10. Some neurons responded to visual or to auditory stimuli in the periphery of the monkey's field of vision.

These neurophysiological findings show that novel, motivational and emotion-provoking visual stimuli influence the activity of neurons in the ventral striatum. These inputs probably reach it from limbic structures such as the amygdala and hippocampus, which project into the ventral striatum and are involved in these motivational, emotional and memory functions (Rolls 1985). In that the majority of these neurons did not have unconditional sensory responses, but instead the response typically depended on memory as to whether the stimulus was recognised, or as to whether it was associated with reinforcement, the function of this part of the striatum does not appear to be purely sensory. Rather, it may provide one route for such memory-related and emotional and motivational stimuli to influence motor output. Thus it will be of interest to determine whether there are changes in responses to such stimuli in Parkinson's disease, or in monkeys with selective depletion of dopamine in the ventral striatum.

Segregation of Function Within the Striatum

These neurophysiological investigations indicate first that there are differences between neuronal activity in different regions of the striatum, and second that the inputs which activate these neurons are derived functionally from the cortical region or limbic structure which projects into each region of the striatum. Thus the majority of neurons in the main part of the putamen (see Rolls et al. 1984) had responses related to movements made by the monkey (see also DeLong et al. 1984). This is consistent with the inputs to these regions from sensorimotor cortex, areas 3,1,2,4 and 6 (see p. 124). In contrast, though the same testing methods were used, neuronal activity related to visual stimuli and which showed rapid habituation was found in the tail of the caudate nucleus and adjoining part of the ventral putamen, which receive from the inferior temporal visual cortex

(Caan et al. 1984). Also, neuronal activity related to the preparation for the initiation of behavioural responses in response to environmental cues was found in the head of the caudate nucleus (Rolls et al. 1979; Rolls et al. 1983), whereas such neurons were relatively rare (10%) in the putamen, and instead neurons with activity unconditionally associated with movements made in the same test situations were common (29%) in the putamen. This is again consistent with the inputs to the head of the caudate nucleus. This region of the striatum receives projections from the prefrontal cortex, in which neurons which respond to the same environmental cues are found (experiments of E. T. Rolls and G. C. Baylis, 1984). Further, the activity of some neurons in the ventral striatum, which receives inputs from limbic structures such as the amygdala and hippocampus, occurs to stimuli related to emotional and novel environmental events which are probably processed through limbic structures (Rolls 1986a, 1987a; Rolls and Williams 1987).

Issues raised by these findings are whether within the striatum there is the possibility for different regions to interact, and whether the partial functional segregation seen within the striatum is maintained in processing beyond the striatum. For example, is the segregation maintained throughout the globus pallidus and thalamus with projections to different premotor and even prefrontal regions reached by different regions of the striatum, or is there convergence or the possibility for interaction at some stage during this post-striatal processing? These questions are considered in the next two sections. It may be noted that although in the other nuclei of the basal ganglia, such as the globus pallidus, substantia nigra pars reticulata, and subthalamic nucleus, there are neurons with activity which is clearly related to leg, arm, or orofacial movements, and there is a somatotopic representation of these body parts within each nucleus (DeLong 1971; DeLong et al. 1983, 1984), the degree to which information other than in movement-related form is present in these nuclei is not yet fully clear. For example, it is not clear whether neurons in the rostrodorsal part of the internal pallidal segment, which receives from the head of the caudate nucleus, respond in relation to environmental cues or to movements. This information will be important in determining whether segregation of function is maintained throughout the basal ganglia.

The Nature of Processing Within the Striatum

Given this evidence for at least partial segregation of types of neuronal response in different regions of the striatum, with neuronal activity in some regions related to movements, but in other regions to more complex events, it may be asked whether some of these latter regions might not have sensory or cognitive functions. Indeed, the experiment of Divac et al. (1967) in which lesions to different parts of the caudate nucleus produced different sensory or cognitive deficits related to the function of the connected cortical region (see p. 126) provides some support for this possibility. However, it is possible to ask whether these parts of the striatum are actually performing a sensory computation, or a cognitive computation (for the present purposes one in which neither the inputs nor the outputs are directly related to sensory or motor function; see also Oberg and Divac 1979), or whether these parts of the striatum provide an essential output route for a cortical area with a sensory or cognitive function but do not themselves have sensory or cognitive functions.

One way to obtain evidence on this is to analyse neurophysiologically the computation being performed by a part of the striatum, and relate this to the computation being performed in its input and output regions. One part of the striatum for which such evidence is available is the head of the caudate nucleus, in which neuronal activity can be compared with neuronal activity in the overlying prefrontal cortex. As described above, information necessary for the computation that a visual stimulus is no longer associated with taste reward reaches the orbitofrontal cortex, and the putative output of such a computation, namely neurons which respond in this non-reward situation, are found in the orbitofrontal cortex (Thorpe et al. 1983). However, such neurons which represent the necessary sensory information for this computation, and neurons which respond to the non-reward, were not found in the head of the caudate nucleus (Rolls et al. 1983). Instead, in the head of the caudate nucleus, neurons in the same test situation responded in relation to whether the monkey had to make a response on a particular trial, that is many of them responded more on Go than on No-Go trials. This could represent the output of a cognitive computation performed by the orbitofrontal cortex, indicating whether, on the basis of the available sensory information, the current trial should be a Go trial, or a No-Go trial because a visual stimulus previously associated with punishment had been shown.

A similar comparison can be made for the tail of the caudate nucleus. Here the visual responses shown by neurons typically habituated to zero within a few trials, whereas such marked habituation was less common in neurons in the inferior temporal visual cortex, which projects to the tail of the caudate nucleus (see Rolls 1987a). In this case, the signal being processed by the striatum thus occurred when a patterned visual stimulus changed, and this could be of use in switching attention or orienting to the changed stimulus.

In both these parts of the striatum in which a comparison can be made of processing in the striatum with that in the cortical area which projects to that part of the striatum, it appears that the full information represented in the cortex does not reach the striatum, but that rather the striatum receives the output of the computation being performed by a cortical area, and could use this to switch or alter behaviour. Thus, if the orbitofrontal cortex computed that a visual signal was no longer associated with reward, it might send the output of this computation to the striatum to indicate that behaviour should be switched from the non-rewarded stimulus. Similarly, a signal from the inferior temporal cortex might indicate that behaviour should be switched to a new patterned stimulus. Comparably, the cue-related neurons in the head of the caudate nucleus which respond to significant environmental events such as a tone cue which precedes the onset of a trial (see p. 127) may reflect an output from the dorsolateral prefrontal cortex which may be involved in this computation and in which similar neurons are found. The head of the caudate nucleus would thus again be in receipt of a signal decoded by the cortex to which it might be appropriate to switch or initiate behaviour. The processing being performed by the head of the caudate might again in this case not be described as sensory, in that neuronal reponses occur only to significant environmental events, requiring a look-up of the stimulus in association memory to determine whether there were any consequences associated with its presentation previously. Further, the processing being performed in the ventral striatum is in many cases not just sensory, in that, for example, many of the neurons which responded to visual inputs did so preferentially on the basis of whether the stimuli were recognised, or were associated with reinforcement (see responses in groups 1 and 2 in Table 15.1). Much of the sensory and memory-related processing required to determine whether a stimulus is a face, is recognised, or is associated with reinforcement has been performed in and is evident in neuronal responses in structures such as the amygdala (Leonard et al. 1985; Rolls 1985, 1987a), orbitofrontal cortex (Thorpe et al. 1983) and hippocampus (Rolls 1985, 1987a). Although the evidence is thus only starting to become available, the possibility arises from these findings that some parts of the striatum, particularly the caudate nucleus and the ventral striatum, receive the output of these memory-related and cognitive computations, but do not themselves perform them, and instead are involved in switching behaviour as appropriate, as determined by the different, sometimes conflicting, information received. Thus, on this view, the striatum would be particularly involved in the selection of behavioural responses, and in producing one coherent stream of behavioural output, with the possibility to switch if a higher priority input was received. Dopamine could play an important role in setting the sensitivity of this response-selection function, as discussed elsewhere (Rolls et al. 1984; Rolls and Williams 1987).

What Computations are Performed by the Basal Ganglia?

On the hypothesis just raised, different regions of the striatum, or at least the outputs of such regions, would need to interact. Is there within the striatum the possibility for different regions to interact, and is the partial functional segregation seen within the striatum maintained in processing beyond the striatum? For example, is the segregation maintained throughout the globus pallidus and thalamus with projections to different premotor and even prefrontal regions reached by different regions of the striatum, or is there convergence or the possibility for interaction at some stage during this post-striatal processing?

Given the anatomy of the basal ganglia, interactions between signals reaching the basal ganglia could happen in a number of different ways. One would be for each part of the striatum to receive at least some input from a number of different cortical regions. There is evidence for patches of input from different sources to be brought adjacent to each other in the striatum (Van Hoesen et al. 1981; Seleman and Goldman-Rakic 1985; Malach and Graybiel 1987). For example, in the caudate nucleus, different regions of association cortex project to adjacent longitudinal strips (Seleman and Goldman-Rakic 1985). Now, the dendrites of striatal neurons have the shape of large plates which lie at right angles to the incoming cortico-striate fibres (Percheron et al. 1984a,b). Thus one way in which interaction may start in the basal ganglia is by virtue of the same striatal neuron receiving inputs on its dendrites from more than just a limited area of the cerebral cortex. This convergence may provide a first level of integration over limited sets of cortico-striatal fibres. The computation which could be performed by this architecture is discussed below for the inputs to the globus pallidus, where the connectivity pattern is comparable (Percheron et al. 1984a,b). The regional segregation of neuronal response types in the striatum described above is consistent with mainly local integration over limited, adjacent, sets of cortico-striatal inputs, as suggested by this anatomy. Short-range integration or interactions within the striatum may also be produced by the short length (e.g. 0.5 mm) of the intrastriatal axons of striatal neurons. These could produce a more widespread influence if the effect of a strong input to one part of the striatum spread like a lateral inhibition signal. Such a mechanism could contribute to behavioural response selection in the face of different competing input signals to the striatum. Another possible mechanism for interaction within the striatum is provided by the dopaminergic pathway, through which a signal which has descended from, for example, the ventral striatum might influence other parts of the striatum (Nauta and Domesick 1978). Because of the slow conduction speed of the dopaminergic neurons, this system would probably not be suitable for rapid switching of behaviour, but only for more tonic, long-term, adjustments of sensitivity.

Further levels for integration within the basal ganglia are provided by the striato-pallidal and striato-nigral projections (Percheron et al. 1984a,b) (Fig. 15.3). The afferent fibres from the striatum again cross at right angles a flat plate or disc formed by the dendrites of the pallidal or nigral neurons. The discs are approximately 1.5 mm in diameter and are stacked up one upon the next at right angles to the incoming striatal fibres. The dendritic discs are so large that in the monkey there is room for only perhaps 50 such discs which do not overlap in the external pallidal segment, for 10 non-overlapping discs in the medial pallidal

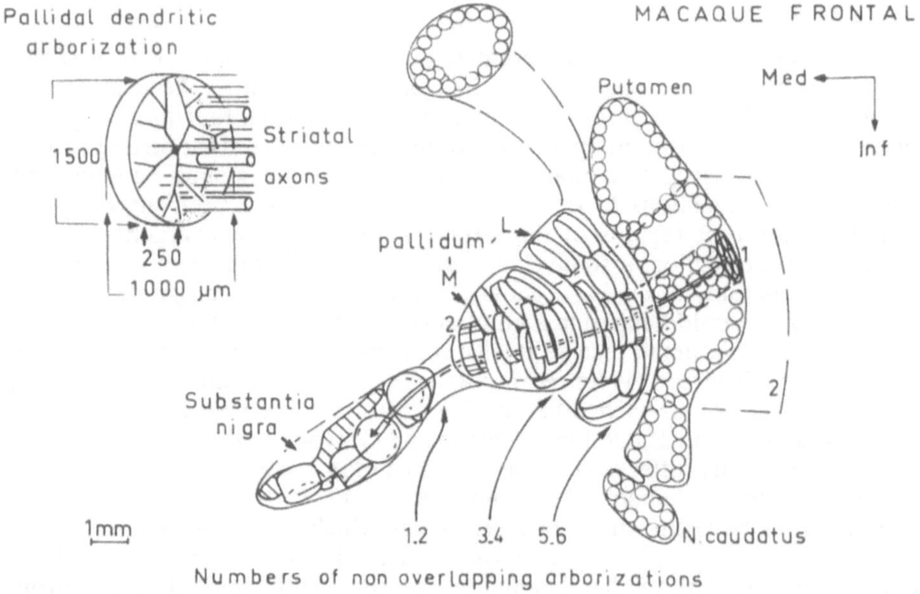

Fig. 15.3. Semi-schematic spatial diagram of the striato-pallido-nigral system (see text). The numbers represent the numbers of non-overlapping arborisations of dendrites in the plane shown. (Reproduced from Percheron et al. 1984b, with permission of the publishers, Plenum Press, New York)

segment, and for one overlapping disc in the most medial part of the medial segment of the globus pallidus and in the substantia nigra (Fig. 15.3).

This arrangement clearly provides a possibility for some convergence onto single pallidal and nigral neurons of signals from relatively different parts of the striatum. For what computation might such anatomy provide a structural basis? Now, each dendritic disc is flat, is orthogonal to the input fibres which pierce it, but is not filled with dendritic arborisations. Instead, each dendrite typically consists of four to five branches which are spread out to occupy only a small part of the surface area of the dendritic disc. There are thousands of such sparsely populated plates stacked on top of one another. Each pallidal neuron is contacted, apparently at random, by a number of the mass of fibres from the striatum which pass it, and given the relatively small collecting area of each pallidal or nigral neuron, each such neuron will thus receive a random combination of inputs from different striatal neurons within its collection field. The thinness of the dendritic sheet may help to ensure that each axon does not make more than a few synapses with each dendrite, and that the combinations of inputs received by each dendrite are approximately random. This architecture thus appears to be appropriate for bringing together at random onto single pallidal and nigral neurons inputs which originate from quite diverse parts of the cerebral cortex. By the stage of the medial pallidum and substantia nigra, there is the opportunity for the input field of a single neuron to effectively become very wide, although whether this covers the whole of the cerebral cortex, or just large parts of it, remains to be determined.

Given then that this architecture allows individual pallidal and nigral neurons to receive random combinations of inputs from different striatal neurons, the following functional implications arise. If a particular pallidal or nigral neuron received inputs by chance from striatal neurons which responded to an environmental cue signal that something significant was about to happen, and from striatal neurons which fired because the monkey was making a postural adjustment, then this conjunction of events might make that pallidal or nigral neuron fire. Such firing produced by conjunctive inputs might then increase the strengths of the synapses from each of the conjunctively active sets of input neurons, according to the well-known Hebb hypothesis of synaptic modification (Rolls 1987a). Then, in the future, occurrence of only one of the inputs, for example only the environmental cue, would result in firing of that neuron, and thus in the appropriate postural adjustment being made by virtue of the output connections of that pallidal or nigral neuron.

Thus, the proposal is that the basal ganglia are able to detect combinations of conjunctively active inputs from quite widespread regions of the cerebral cortex using their combinatorial architecture and a property of synaptic modifiability. In this way it would be possible to trigger any complex pattern of behavioural responses by any complex pattern of environmental inputs, using what is effectively a random associative net, the properties of which are described elsewhere (Rolls 1987a). It may be noted that the input events need not include only those from environmental stimuli represented in the caudate nucleus and ventral striatum, but also, if the overlapping properties of the dendrites described above provide sufficient opportunity for convergence, of the context of the movement also, provided by inputs via the putamen from sensorimotor cortex. This would then make a system appropriate for triggering an appropriate motor response (learned by trial and error, with the final solution becoming associated with the triggering input events) to any environmental input state. As such, the hypothesis could be said to provide a basis for the storage of motor plans in the basal ganglia, which would be instantiated as a series of look-ups of the appropriate motor output pattern to an evolving sequence of input information. The hypothesis may also provide a basis for the switching between different types of behaviour proposed as a function of the basal ganglia, for if a strong new pattern of inputs was received by the basal ganglia, this would result in a different pattern of outputs being "looked up" than that currently in progress.

Evidence relevant to determining the utility of this hypothesis will include further information on the extent to which convergence from widely separate parts of the cerebral cortex is achieved in the basal ganglia, and, if not, on whether it is achieved in the thalamic or cortical structures to which the basal ganglia project (see p. 124); and evidence on whether pallidal and nigral neurons show learning of conjunctions of their inputs. The questions of whether the ventral pallidum and the globus pallidus allow for any form of convergence; of whether there are separate regions of the globus pallidus, ventral pallidum, or substantia nigra which do not contain movement-related neurons but in which neuronal activity is similar to that in the head of the caudate nucleus and ventral pallidum; and of the exact role of the outputs to different premotor and prefrontal regions from different parts of the basal ganglia (see Figs. 15.1, 15.2) are also of importance. It may be noted that one reason for having two stages of processing in the basal ganglia, from the cortex to striatal neurons, and from striatal neurons to pallidal or nigral neurons, may be to ensure that the proportion of responding input neurons to any stage of the memory is kept relatively low, in order to reduce interference in the association memory (Rolls 1987a).

An alternative view of striatal function is that the striatum might be organised as a set of segregated and independent transmission routes, each one of which would receive from a given region of the cortex, and project finally to separate premotor or prefrontal regions (see Fig. 15.2). On this view, detection of combinations of conjunctively active inputs, but in this case from limited populations of input axons to the basal ganglia, might still be an important aspect of the function of the basal ganglia. More investigations are needed to lead to further understanding of these conceptual ideas on the function of the basal ganglia.

Disturbances of Striatal Function

The findings described here on neuronal activity in different regions of the striatum have implications for our understanding of striatal dysfunction. If the transmission of information from the cortex through the striatum to its outputs was decreased, by, for example, depletion of dopamine, then the disturbance of behaviour which resulted would not represent just a movement disorder. For example, the akinesia of Parkinson's disease could be partly due to an inability to respond to environmental cues normally used in the preparation for and initiation of movement. This function normally engages neurons in the head of the caudate nucleus. Similarly, the sensorimotor deficit produced by damage to the nigro-striatal bundle could occur partly because neurons in the tail of the caudate nucleus which normally respond to changing patterned visual stimuli are insensitive to these stimuli. Further, some alteration in emotional and motivational responses might arise if the functions of the ventral striatum were disturbed by depletion of dopamine. Finally, the movement disorders of Parkinson's disease, such as the inability to grade the amplitude of a movement correctly, might be related to disturbance of the function of the putamen, in which neurons are found which respond in relation to the amplitude of movements. It may be emphasised that the relative magnitude of these symptoms might be different in different patients, depending on the degree of depletion of dopamine in different regions of the striatum.

Investigations of whether there are cognitive deficits in Parkinson's disease are now in progress. For example, there is a difficulty in shifting conceptual sets, and more perseverative errors are produced on both a modified Wisconsin card sorting task and Benton's word fluency test (Lees and Smith 1983). However, whether such deficits reflect altered function of the striatum or of the prefrontal cortex (in which dopamine is also depleted; Scatton et al. 1983) is difficult to establish.

According to this analysis, the opposite type of functional disorder, in which transmission through the striatum was abnormally elevated, might lead (1) to increased responsiveness to temporal lobe signals, with failure of these to habituate normally, so leading perhaps to hallucinations; and (2) to increased reactivity to environmental signals, even perhaps when these were irrelevant, so that distractability, inability to maintain attention and concentrate on one behaviour, and an increased tendency to switch between behaviours might result. Moreover, because there is some segregation of function within the striatum, and some topography of the dopamine projection, these symptoms might be dissociable. Thus in some cases cognitive dysfunctions as compared with motor dysfunctions might predominate. It is thus possible that some of the symptoms of schizophrenia could be related to over-responsiveness of these striatal systems. In line with this hypothesis, increased dopamine receptor binding, implying increased sensitivity to dopamine, has been reported in the striatum of some schizophrenic patients (even when they have not been treated with neuroleptic drugs which might lead to supersensitivity; Lee et al. 1978; Matthyse 1981). Further, the neuroleptic drugs used to treat schizophrenia do block dopamine receptors (Matthyse 1981), and might thus be expected to normalise some of the over-responsiveness described above (Crow 1979). Although this parallel between neuronal activity in the striatum and some of the symptoms of schizophrenia is of interest, there are also dopamine projections to the prefrontal cortex and amygdala, and it is unlikely that altered neuronal responsiveness in the striatum plays an exclusive role in the symptoms observed.

Conclusion

In this chapter it has been shown (1) that the striatum receives inputs from different areas of the cerebral cortex, and allows these to influence premotor and prefrontal cortical areas, and (2) that there is considerable segregation of function within the striatum. It is suggested (3) that there is an opportunity for inputs from different parts of the cerebral cortex to interact on the dendrites of single neurons in the globus pallidus and substantia nigra, and (4) that these parts of the basal ganglia may learn associations between the different signals received from the striatum, so that the basal ganglia provide a way for cortical areas far on in the hierarchy of information processing to execute particular sequences of movements, and thus to execute motor programs.

References

Anderson ME (1978) Discharge patterns of basal ganglia neurons during active maintenance of postural stability and adjustment to chair tilt. Brain Res 143:325–338

Baylis GC, Rolls ET, Leonard CM (1985) Selectivity between faces in the responses of a population of neurons in the cortex in the superior temporal sulcus of the monkey. Brain Res 342:91–102

Baylis GC, Rolls ET, Leonard CM (1987) Functional subdivisions of temporal lobe neocortex. J Neurosci 7:330–342

Buerger AA, Gross CG, Rocha-Miranda CE (1974) Effects of ventral putamen lesions on discrimination learning by monkeys. J Comp Physiol Psychol 86:440–446

Caan W, Perrett DI, Rolls ET (1984) Responses of striatal neurons in the behaving monkey. 2. Visual processing in the caudal neostriatum. Brain Res 290:53–65

Carpenter MB (1984) Interconnections between the corpus striatum and brain stem nuclei. In: McKenzie JS, Kemm RE, Wilcock LN (eds) The basal ganglia. Structure and function. Plenum, New York, pp 1–68

Crow TJ (1979) What is wrong with dopaminergic transmission in schizophrenia? Trends Neurosci 2:52–55

Crutcher MD, DeLong MR (1984a) Single cell studies of the primate putamen. I. Functional organization. Exp Brain Res 53:233–243

Crutcher MD, DeLong MR (1984b) Single cell studies of the primate putamen. II. Relations to direction of movements and pattern of muscular activity. Exp Brain Res 53:244–258

DeLong MR (1971) Activity of pallidal neurons during movement. J Neurophysiol 34:414–427

DeLong MR, Crutcher MD, Georgopoulos AP (1983) Relations between movement and single cell discharge in the substantia nigra of the behaving monkey. J Neurosci 3: 1599–1606

DeLong MR, Georgopoulos AP, Crutcher MD, Mitchell SJ, Richardson RT, Alexander GE (1984) Functional organization of the basal ganglia: contributions of single-cell recording studies. In: Functions of the basal ganglia. Ciba Foundation Symposium 107. Pitman, London, pp 64–78

Divac I, Oberg RGE (1979) Current conceptions of neostriatal functions. In: Divac I, Oberg RGE (eds) The neostriatum. Pergamon, New York, pp 215–230

Divac I, Rosvold HE, Szwarcbart MK (1967) Behavioral effects of selective ablation of the caudate nucleus. J Comp Physiol Psychol 63:184–190

Dunnett SB, Iversen SD (1981) Learning impairments following selective kainic acid-induced lesions within the neostriatum of rats. Behav Brain Res 2:189–209

Dunnett SB, Iversen SD (1982a) Neurotoxic lesions of ventrolateral but not anteromedial neostriatum impair differential reinforcement of low rates (DRL) performance. Behav Brain Res 6:213–226

Dunnett SB, Iversen SD (1982b) Sensorimotor impairments following localised kainic acid and 6-hydroxydopamine lesions of the neostriatum. Brain Res 248:121–127

Evarts EV, Wise SP (1984) Basal ganglia outputs and motor control. In: Functions of the basal ganglia. Ciba Foundation Symposium 107. Pitman, London, pp 83–96

Gaffori O, LeMoal M (1979) Disruption of maternal behaviour and appearance of cannabalism after ventral mesencephalic tegmentum lesions. Physiol Behav 23:317–322

Heimer L, Wilson RD (1975) The subcortical projections of the allocortex: similarities in the neuronal associations of the hippocampus, the pyriform cortex and the neocortex. In: Santini M (ed) Golgi centennial symposium: perspectives in neurobiology. Raven, New York, pp 177–193

Heimer L, Switzer RD, Van Hoesen GW (1982) Ventral striatum and ventral pallidum. Additional components of the motor system? Trends Neurosci 5:83–87
Hemphill M, Holm G, Crutcher M, Delong M, Hedreen J (1981) Afferent connections of the nucleus accumbens in the monkey. In: Chronister RB, DeFrance JF (eds) The neurobiology of the nucleus accumbens. Haer: Brunswick, New Jersey, pp 75–81
Hornykiewicz O (1973) Dopamine in the basal ganglia: its role and therapeutic implications. Br Med Bull 29:172–178
Iversen SD (1979) Behaviour after neostriatal lesions in animals. In: Divac I, Oberg RGE (eds) The neostriatum. Pergamon, Oxford, pp 195–210
Iversen SD (1984) Behavioural effects of manipulation of basal ganglia neurotransmitters. In: Functions of the basal ganglia. Ciba Foundation Symposium 107. Pitman, London, pp 183–195
Jones EG, Coulter JD, Burton H, Porter R (1977) Cells of origin and terminal distribution of corticostriatal fibres arising in sensory motor cortex of monkeys. J Comp Neurol 181:53–80
Jurgens U (1984) The efferent and afferent connections of the supplementary motor area. Brain Res 300:63–81
Kemp JM, Powell TPS (1970) The cortico-striate projections in the monkey. Brain 93:525–546
Kemp JM, Powell TPS (1971) The connections of the striatum and globus pallidus: synthesis and speculation. Philos Trans Soc Lond [Biol] 262: 441–457
Kunzle H (1975) Bilateral projections from precentral motor cortex to the putamen and other parts of the basal ganglia. Brain Res 88:195–209
Kunzle H (1977) Projections from primary somatosensory cortex to basal ganglia and thalamus in the monkey. Exp Brain Res 30:481–482
Kunzle H (1978) An autoradiographic analysis of the efferent connections from premotor and adjacent prefrontal regions (areas 6 and 9) in Macaca fascicularis. Brain Behav Evol 15:185–234
Kunzle H, Akert K (1977) Efferent connections of area 8 (frontal eye field) in Macaca fascicularis. J Comp Neurol 173:147–164
Lee T, Seeman P, Tourtelotte WW, Farley IJ, Hornykiewicz O (1978) Binding of 3H neuroleptics and 3H-apomorphine in schizophrenic brains. Nature 274:897–900
Lees AJ, Smith E (1983) Cognitive deficits in the early stages of Parkinson's disease. Brain 106:257–270
Leonard CM, Rolls ET, Wilson FAW, Baylis GC (1985) Neurons in the amygdala of the monkey with responses selective for faces. Behav Brain Res 15:159–176
Malach R, Graybiel AM (1987) The somatic sensory cortico-striatal projection: patchwork of somatic sensory zones in the extra-striasomal matrix. In: Schneider JS, Lidsky TI (eds) Basal ganglia and behavior. Huber, New York, pp 11–16
Marshall JP, Richardson JS, Teitelbaum P (1974) Nigrostriatal bundle damage and the lateral hypothalamic syndrome. J Comp Physiol Psychol 87:808–830
Matthyse S (1981) Nucleus accumbens and schizophrenia, 1980. In: Chronister RB, DeFrance JF (eds) The neurobiology of the nucleus accumbens. Haer, Brunswick, New Jersey, pp 351–359
Nauta WJH, Domesick VB (1978) Crossroads of limbic and striatal circuitry: hypothalamonigral connections. In: Livingstone KE, Hornykiewicz O (eds) Limbic mechanisms. Plenum, New York, pp 75–93
Nauta WJH, Domesick VB (1984) Afferent and efferent relationships of the basal ganglia. In: Functions of the basal ganglia. Ciba Foundation Symposium 107. Pitman, London, pp 3–29
Newman R, Winans SS (1980a) An experimental study of the ventral striatum of the golden hamster. I. Neuronal connections of the nucleus accumbens. J Comp Neurol 191:167–192
Newman R, Winans SS (1980b) An experimental study of the ventral striatum of the golden hamster. II. Neuronal connections of the olfactory tubercle. J Comp Neurol 191:193–212
Oberg RGE, Divac I (1979) "Cognitive" functions of the neostriatum. In: Divac I, Oberg RGE (eds) The neostriatum. Pergamon, New York, pp 291–313
Percheron G, Yelnik J, Francois C (1984a) A Golgi analysis of the primate globus pallidus. III. Spatial organization of the striato-pallidal complex. J Comp Neurol 227:214–227
Percheron G, Yelnik J, Francois C (1984b) The primate striato-pallido-nigral system: an integrative system for cortical information. In: McKenzie JS, Kemm RE, Wilcox LN (eds) The basal ganglia: structure and function. Plenum, New York, pp 87–105
Rolls ET (1984a) Neurons in the cortex of the temporal lobe and in the amygdala of the monkey with responses selective for faces. Hum Neurobiol 3:209–222
Rolls ET (1984b) Activity of neurons in different regions of the striatum of the monkey. In: McKenzie JS, Kemm RE, Wilcox LN (eds) The basal ganglia: structure and function. Plenum, New York, pp 467–493

Rolls ET (1985) Connections, functions and dysfunctions of limbic structures, the prefrontal cortex, and hypothalamus. In: Swash M, Kennard C (eds) The scientific basis of clinical neurology. Churchill Livingstone, London, pp 201–213

Rolls ET (1986a) A theory of emotion, and its application to understanding the neural basis of emotion. In: Oomura Y (ed) Emotions. Neural and chemical control. Japan Scientific Societies Press, Tokyo; Karger, Basel, pp 325–344

Rolls ET (1986b) Neuronal activity related to the control of feeding. In: Ritter RC, Ritter S, Barnes CD (eds) Feeding behaviour: neural and humoral controls. Academic, New York, pp 163–190

Rolls ET (1987a) Information representation, processing and storage in the brain: analysis at the single neuron level. In: Changeux J-P, Konishi M (eds) Neural and molecular mechanisms of learning, Wiley, Chichester

Rolls ET (1987b) A neurophysiological systems approach to neuroethology. In: Guthrie DM (ed) Aims and methods in neuroethology. Manchester University Press, Manchester, pp 231–239

Rolls ET (1988) Visual information processing in the primate temporal lobe. In: Imbert M (ed) Models of visual perception: from natural to artificial, Oxford University Press, Oxford (in press)

Rolls ET, Williams GV (1987) Sensory and movement-related neuronal activity in different regions of primate striatum. In: Schneider JS, Lidsky TI (eds) Basal ganglia and behavior. Huber, New York, pp 37–59

Rolls ET, Thorpe SJ, Maddison S, Roper-Hall A, Puerto A, Perrett D (1979) Activity of neurones in the neostriatum and related structures in the alert animal. In Divac I, Oberg RGE (eds) The neostriatum. Pergamon, Oxford, pp 163–182

Rolls ET, Ashton J, Williams G, Thorpe SJ, Mogenson GJ, Colpaert F, Phillips AG (1982) Neuronal activity in the ventral striatum of the behaving monkey. Soc Neurosci Abstr 8:169

Rolls ET, Thorpe SJ, Maddison J (1983) Responses of striatal neurons in the behaving monkey. 1. Head of the caudate nucleus. Behav Brain Res 7:179–210

Rolls ET, Thorpe SJ, Boytim M, Szabo I, Perrett DI (1984) Responses of striatal neurons in the behaving monkey. 3. Effects of iontophoretically applied dopamine on normal responsiveness. Neuroscience 12:1201–1212

Sanghera MK, Rolls ET Roper-Hall A (1979) Visual responses of neurons in the dorsolateral amygdala of the alert monkey. Exp Neurol 63:610–626

Scatton B, Javoy-Agid F, Rouquier L, Dubois B, Agid Y (1983) Reduction of cortical dopamine, noradrenaline, serotonin and their metabolites in Parkinson's disease. Brain Res 275:321–328

Schell GR, Strick PL (1984) The origin of thalamic inputs to the arcuate premotor and supplementary motor areas. J Neurosci 4:539–560

Seleman LD, Goldman-Rakic PS (1985) Longitudinal topography and interdigitation of corticostriatal projections in the rhesus monkey. J Neurosci 5:776–794

Stevens JR (1979) Schizophrenia and dopamine regulation in the mesolimbic system. Trends Neurosci 2:103–105

Stinus L, Gaffori O, Simon H, LeMoal M (1978) Disappearance of hoarding and disorganisation of eating behaviour after ventral mesencephalic tegmentum lesions in rats. J Comp Physiol Psychol 92:289–329

Thorpe SJ, Rolls ET, Maddison SP (1983) Neuronal responses in the orbitofrontal cortex of the behaving monkey. Exp Brain Res 49:93–115

Tobias TJ (1975) Afferents to the prefrontal cortex from the thalamic mediodorsal nucleus in the rhesus monkey. Brain Res 83:191–212

Van Hoesen GW, Mesulam M-M, Haaxma R (1976) Temporal cortical projections to the olfactory tubercle in the rhesus monkey. Brain Res 109:375–381

Van Hoesen GW, Yeterian EH, Lavizzo-Mourey R (1981) Widespread corticostriate projections from temporal cortex of the rhesus monkey. J Comp Neurol 199:205–219

Webster (1988) In: Stern G (ed) Parkinson's disease. Chapman and Hall, London (in press)

Whitlock DG, Nauta WJH (1956) Subcortical projections from the temporal neocortex in Macaca mulatta. J Comp Neurol 106:183–212

Yeterian EH, Van Hoesen GW (1978) Cortico-striate projections in the rhesus monkey: the organization of certain cortico-caudate connections. Brain Res 139:43–63

Chapter 16

Muscle Afferents and Parkinson's Disease

P. B. C. Matthews

It is clear that Hughlings Jackson was entirely familiar with Parkinson's disease, although he wrote very little about it. He did, however, give us the essential idea that the tremor and rigidity are released symptoms resulting from the abnormal activity of the lower centres, rather than due to a direct action of the higher centres themselves. Let me quote what he said at a discussion meeting in 1888, when he was much more direct than in his more prepared writings. He stated, "I have submitted the hypothesis, that in paralysis agitans there is wasting of the cells of the middle motor centres . . . such a process, a negative one, can cause only the negative symptom paralysis. But being at the same time a loss of control over the anterior horns, there is over activity", which he saw as resulting from "the taking off of inhibition from the anterior horns" so causing, in his view, first tremor then rigidity (Jackson 1888).

The story was taken up in the next generation by his apostle, Sir Francis Walshe. Walshe, like Jackson, spent his life immersed in busy clinical practice, and he also wrote widely. Much of this was expounding and developing Jackson's views for those who found them hard to grasp in the original. It is rather as if Jackson had followed his initial leanings to become a philosopher, and had become an earlier Wittgenstein requiring interpretation for the multitude. The torch passed directly to Walshe from Jackson. In his notes for his subsequent biographer Walshe wrote, "familiar figures of my youth were Sir William Gowers and Dr. Hughlings-Jackson, who lived within a few hundred yards of my home, whose work and scientific evidence were constantly impressed on me by my father – this determining in large measure a later leaning to specialise in clinical neurology."

In 1924, at the outset of his career, Walshe himself performed experiments on Parkinsonian patients. He injected local anaesthetic into the rigid muscles, and produced a melting away of the rigidity at a time when the patient retained full muscle power, so the motor fibres themselves were not paralysed. Walshe, following the animal work of that time, attributed this to a selective paralysis of the afferent fibres to muscle (Walshe 1924). With hindsight it is more likely that he was paralysing the gamma efferents, the motor fibres to the muscle spindles; but that is very much the same thing for us, since he produced a partial functional de-afferentation and this was what abolished the rigidity. Tremor was unaffected by the injection and, what is much more interesting, he found that voluntary performance was improved, thereby suggesting that the rigidity was not just a sign for the neurologist to observe, but was also somehow interfering with the patient's activity. His striking finding was that the patient experienced a new freedom of action when employing the now flaccid muscles, and a variety of movements which had long been impossible were transiently restored. Thus "one patient was able for the first time in eighteen months to raise a cup to his lips and drink unassisted after biceps and triceps" had been injected. So we have the challenging conclusion that both the symptoms and the signs of Parkinsonism owe a great deal to the afferent activity coming from the rigid muscles. The rest of this chapter is concerned with rigidity; the equally interesting question of the genesis of the tremor has recently been taken up by Rack and Ross (1986).

There are two initially tempting hypotheses as to the nature of the abnormality underlying rigidity, but neither of them can continue to be held. The first is that there is a heightened responsiveness somewhere in the monosynaptic reflex arc from the muscle spindle primary endings to the motor neurons. The knock-down argument against this is simply the classic observation that tendon jerks are not changed in Parkinsonism, which was probably known to Jackson and was certainly referred to by Walshe. The second more recent idea is that there might be fusimotor overactivity, with the muscle spindles being driven to fire too rapidly and so causing an abnormally large stretch reflex. This will not do either, because the microneurographic recordings by Burke et al. (1977) show that the spindle firing is very much the same in Parkinsonian patients as in the normal, provided they are studied under equivalent conditions with the muscle contracting in the normal as well as in the Parkinsonian patient.

This leaves us with a third suggestion, that of Tatton and Lee of some 10 years ago (Tatton and Lee 1975; Lee and Tatton 1978), that there is overactivity in a special long-latency stretch reflex pathway. This idea is illustrated in Fig. 16.1.

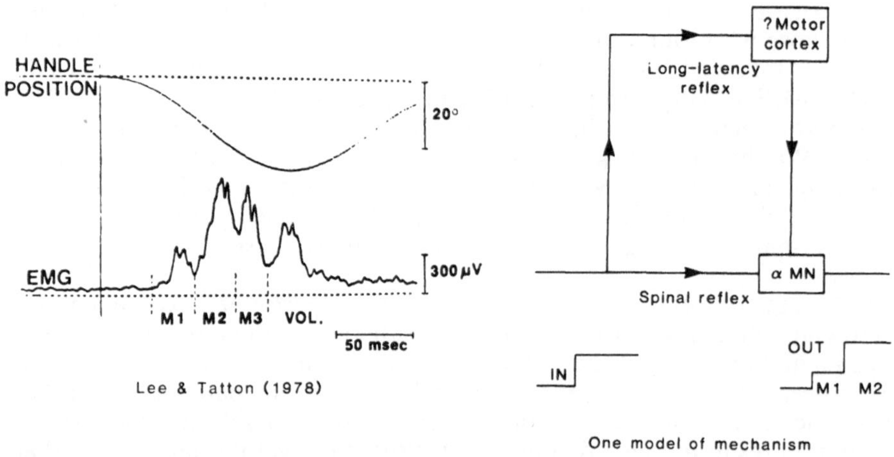

Fig. 16.1. *Left*, an example of the complex electromyographic response evoked from a muscle on stretching it by displacing the joint at which it acts. This example is for the wrist extensors. *VOL*, voluntary. *Right*, schema illustrating how the later M2/M3 wave might arise from the activation of a "long-loop" reflex involving the cortex. (Reproduced from Matthews 1985, with permission of the publishers, Croom Helm, London)

Normal human subjects grasped a handle, which was then suddenly rotated so as to stretch the flexor muscles in the forearm. Electromyographic recordings from the stretched muscles were rectified and averaged over a number of trials, to give responses like those on the left of Fig. 16.1. The EMG shows a whole series of waves, for which the authors introduced the terminology "M1" for the initial monosynaptic short-latency reflex and "M2/M3" for the later long-latency reflex. At much the same time Marsden et al. (1976) observed a similar late wave for the long flexor of the thumb. Both groups made similar suggestions as to the origin of this later response. They said, in effect, that the same Ia volley from spindle primary endings goes both directly to the alpha motor neurons to elicit the initial M1 response, and also up to the cortex and back again to produce the later waves, as illustrated on the right of Fig. 16.1. So the delay occurs because the same volley takes a circuitous route inside the central nervous system. Tatton and Lee went on to show that the later M2/M3 response was enlarged in some Parkinsonian patients and suggested that this provided a prime candidate for the causation of Parkinsonism rigidity. Moreover, they suggested that the change in reflex excitability was occurring up in the motor cortex with its wealth of connections. However, it is not a change in the excitability of the pyramidal tract neuron itself. Berardalli et al. (1984) tested this by the direct method of "shocking" the pyramidal tract cells through the intact skull and finding no change in the Parkinsonian patient. Of course, that finding does not eliminate the possibility that the excitability of the postulated long-loop pathway is increased at some other site.

At this stage we should note that the long-loop hypothesis is not the only possible explanation for the later components of the stretch reflex. Next, there is the view of Eklund et al. (1982) that there is no separate long-latency reflex, but just a continuing short-latency reflex action elicited by the continuing stimulus, with a segmentation into the various waves because of muscle resonances etc. That does not look very plausible for Parkinsonism, because it does not begin to explain how you can have an accentuated later response without an accentuated earlier response.

There is a third possibility, which I am canvassing myself but which is still controversial, that the delayed reflex loses time because it uses the slower afferents coming from the muscle spindle secondary ending (Matthews 1984). Let me remind you that the muscle spindle has two kinds of sensory endings, namely a single primary ending and usually one or more secondary endings. The primary ending is the one that everybody talks about when they are thinking of the spindle. The secondary endings tend to get relegated to the second division, but there are just as many as the primary endings. The important thing for the present argument is that the afferent fibres of the secondary endings are smaller than the afferent fibres of the primary endings and so conduct more slowly. In consequence, if a stimulus sets up a peripheral volley at the same time in the two groups of afferents, the volley from the secondaries will reach the spinal cord appreciably later. In a laboratory animal like the cat, that does not make much difference, since the difference is only a couple of milliseconds. But in man, it takes some 15 ms for the fast Ia volley from the primaries to reach the cord from the forearm, and the slower group II volley from the spindle secondaries will probably take about twice as long. Thus if these slower group II afferents produce reflex excitation it will occur some 15 ms or more after that produced by the fast afferents, as does the long-latency response. Whether they regularly produce excitation is unknown, and the animal work is inconclusive. It is my hypothesis that thcy may do so in intact man and that this is the cause of at least part of the delayed reflex.

Both sets of afferents are excited by muscle stretch. Their actions can be differentiated by using as a stimulus, not simple stretch, but high-frequency vibration. For various reasons, the primary ending is very much more sensitive than the secondary endings to vibration at about 100 Hz, so by using this as a stimulus the afferent volleys are largely from the primary endings, with very little secondary contribution. So, if a reflex is due just to the primary endings it will be produced just as well by vibration. But if it is due to the secondary endings then "genuine" large amplitude stretch will be a much more effective stimulus. In normals, stretch is indeed a far more effective stimulus for the late response than is vibration, which is the basis for my suggestion that the long-latency response is due to the spindle group II afferents. All kinds of controls are needed to make the argument stick and the hypothesis can certainly not be regarded as finally proven. Moreover, even if established it does not exclude the other potential mechanisms (long-loop, segmentation by resonances etc.) from also contributing to the late responses.

It seemed interesting to further test the group II hypothesis by looking at the heightened M2 response of Parkinsonism, to see whether or not the vibration response is also augmented (Cody et al 1986). We recorded the EMG of flexor carpi radialis while the subject was making a weak voluntary wrist flexion and elicited a reflex reponse from it, first by forcible wrist extension and, second, by vibration appled to its tendon. Figure 16.2 shows typical results. With stretch, the Parkinsonian patient shows a normal short-latency response followed by an accentuated long-latency response, as has already been well described by others. With vibration, both the normal subject and the patient show a clear short-latency response followed by a flat baseline (i.e. nothing, no late response). At first sight it might be thought that as the vibration continues so also should the reflex response, because we can be expected to be continuing to bombard the

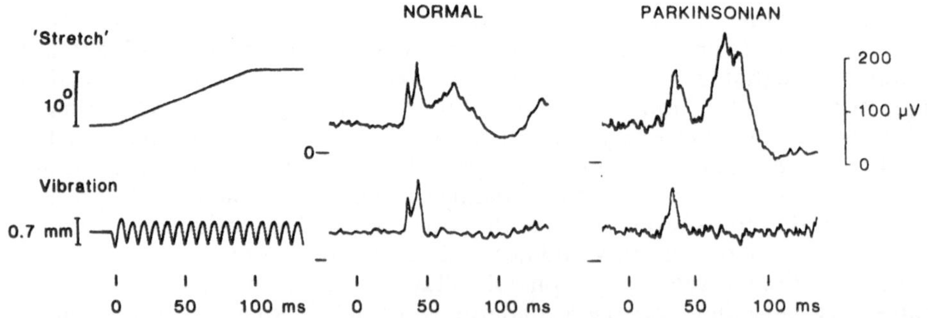

Fig. 16.2. The contrasting EMG responses of flexor carpi radialis to wrist dorsiflexion ('*Stretch*') and to tendon vibration, in a normal subject and in a patient with Parkinson's disease with moderate rigidity. Between the stimuli the subjects maintained a flexor force of about 20% maximal, but they avoided responding voluntarily to the stimuli. Sixty-four trials were averaged in each case. The *upper* stimulus signal shows the angular movement applied to the wrist; the *lower* shows the vibratory movement applied transcutaneously to the flexor tendon a few centimetres above the wrist. The *short horizontal bar* beneath the beginning of each of the rectified EMG records gives its zero level. (Reproduced from Cody et al. 1986, with permission of the Editor of *Brain*)

motor neurons with Ia volleys. It does not go on, I believe, because those motor neurons which are nearly ready to fire, all fire in response to the first Ia volley, and then they are refractory; in addition, they inhibit their neighbours via the Renshaw cells. The late wave evoked by stretch requires an additional excitatory input to the motoneurons, over and above the continuing Ia short-latency action. If the late wave were due to Ia volleys taking a transcortical route, we would expect to see a late wave equally with vibration because it sets up perfectly good Ia volleys as judged by the initial short-latency response. The group II afferents, however, are well placed to provide the delayed input, since they are excited by stretch and they do conduct more slowly. The basic element of hypothesis which is now reiterated for Parkinsonism, is that they produce significant excitation of the motoneurons by a spinal pathway. The essential evidence for this idea is that vibration again produces a pure short-latency response without a long-latency response, in spite of the earlier-suggested increase in the excitability of the cortical relays etc. of the postulated Ia long-loop.

In studying the population as a whole the tendon jerk was not changed systematically with Parkinsonism, nor was the response to vibration. M2, however, showed a clear increase in Parkinsonism, with the mean value significantly greater than the normal. But it was not an absolute separation, and there was appreciable overlap between the patients and the normal subjects. Tatton et al. (1984) suggested that there was no overlap, that there were two separate populations, normal subjects and patients with rigidity. But in line with various other workers (Berardelli et al. 1983; Rothwell et al., 1983) we could not confirm this and found the derangement in Parkinsonism to be statistical rather than absolute. Since only some, not all, rigid Parkinsonian patients have very large M2s, it follows that this cannot be the sole cause of the rigidity. Other factors must also be operating, though it is tempting to believe that, when present, the enhanced long-latency response plays some part in the genesis of the rigidity.

Finally, I want to mention an additional finding which illustrates the complexity of the mechanisms we are dealing with. In the normal subject, the stretch response is often abruptly cut off while the stretch is still going on (including its rising phase, see Fig. 16.1). Something has brought it to an end. Some of this may be refractoriness, but one suspects that an active inhibitory process is largely responsible. On average, we found that in the Parkinsonian patients the M2 wave lasted longer than the normal M2, as illustrated in Fig. 16.3. This suggests that not only may excitatory mechanisms be turned up, but inhibitory mechanisms may be turned down.

What can we conclude from all this?

1. In agreement with everybody else, we find that the long latency response tends to be increased in Parkinsonian patients, but the response is not bimodal; the Parkinsonian patients overlap with the normal subjects. At the very least, therefore, the rigidity must be due to other factors in addition to any increase in the long-latency reflex. It is tempting to believe that this latter plays some part in producing rigidity, but even this remains unproven.

2. The vibration evidence suggests that any accentuation of the long-latency reflex is not a consequence of a change of transcortical excitability to a Ia input, but a change of spinal excitability to a group II input. With Jackson, I suspect that the functional change is taking place at the lowest Jacksonian level, consequent upon its release.

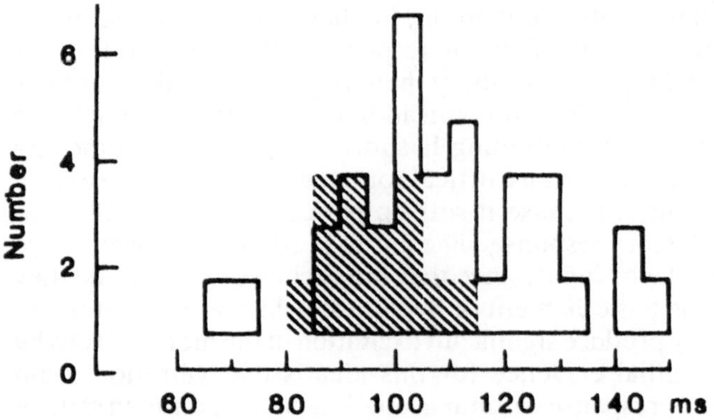

Fig. 16.3. Comparison of the times of termination of the M2 response (from the beginning of stretch) of Parkinsonian patients (*solid lines enclosing clear area*) and of normal subjects (*cross-hatched*). The average duration of the M2 response is significantly longer in the patients, since both groups started at much the same time. (Reproduced from Cody et al. 1986, with permission of the Editor of *Brain*)

3. In looking for release mechanisms we should not just be thinking of excitatory mechanisms, we should be thinking equally of inhibitory mechanisms, and how they have been changed.

4. Finally, a general message for all those studying Parkinsonism: continue to remember Jackson's ideas and Walshe's findings. We should concentrate attention on the lower spinal levels as much as upon the highest levels. We cannot expect to unravel the functional derangement in Parkinsonism by working solely on the basal ganglia themselves. We must come down to the spinal cord as well and understand what is going on there. The goal before us is that deranged spinal mechanisms may turn out to be the immediate cause not only for the rigidity, but also for some of the poverty of movement. Walshe's crucial finding, that reduction of the afferent outflow from muscle allows a patient to move freely again, provides a continuing challenge both for theoretical understanding and for therapeutic advance.

References

Berardelli A, Sabra AF, Hallett M (1983) Physiological mechanisms of rigidity in Parkinson's disease. J Neurol Neurosurg Psychiatry 46:45–53

Berardalli A, Cowan J, Day BL, Dick JPR, Marsden CD, Rothwell JC (1984) Motor cortex excitability in Parkinson's disease. J Physiol 353:33P

Burke D, Hagbarth K-E, Wallin BG (1977) Reflex mechanisms in parkinsonian rigidity. Scand. J. Rehabil Med 9:15–23

Cody FWJ, MacDermott N, Matthews PBC, Richardson HC (1986) Observations on the genesis of the stretch reflex in Parkinson's disease. Brain 109:229–250

Eklund G, Hagbarth K-E, Hagglund JV, Wallin EU (1982) The 'late' reflex responses to muscle stretch: the 'resonance hypothesis' versus the 'long-loop hypothesis'. J Physiol (Lond) 326:79–90

Jackson JH (1888) Discussion on muscular hypertonicity in paralysis. Brain 10:312–318

Lee RG, Tatton WG (1978) Long loop reflexes in man: clinical applications. In: Desmedt JE (ed) Cerebral motor control in man: long loop mechanisms. Karger, Basel, pp 167–177

Marsden CD, Merton PA, Morton HB (1976) Servo action in the human thumb. J Physiol (Lond) 257:1–44
Matthews PBC (1984) Evidence from the use of vibration that the human long-latency stretch reflex depends upon spindle secondary afferents. J Physiol (Lond) 348:383–415
Matthews PBC (1985) Human long-latency stretch reflexes – a new role for the secondary ending of the muscle spindle? In: Barnes WJP, Gladden MH (eds) Feedback and motor control in invertebrates and vertebrates. Croom Helm, London, pp 431–449
Rack PMA, Ross AF (1986) The role of reflexes in the resting tremor of Parkinson's disease. Brain 104:115–141
Rothwell JC, Obeso JA, Traub MM, and Marsden CD (1983) The behaviour of the long-latency stretch reflex in patients with Parkinson's disease. J Neurol Neurosurg Psychiatry 46:35–44
Tatton WG, Lee RG (1975) Evidence for abnormal long-loop reflexes in rigid parkinsonian patients. Brain Res 100:671–676
Tatton WG, Bedingham W, Verrier MC, Blair RDG (1984) Characteristic alterations in responses to imposed wrist displacement in parkinsonian rigidity and dystonia musculorum derformans. Can J Neurol Sci 11:281–287
Walshe FMR (1924) Observations on the nature of the muscular rigidity of paralysis agitans, and on its relationship to tremor. Brain 47:159–177

Chapter 17

Hierarchical Aspects of Eye Movement Disorders

C. Kennard

Although John Hughlings Jackson often referred to ocular motor control in relation to his general hypothesis of the hierarchical control of movement, he appears only to have written one paper specifically on the subject. It was written in 1909, and was entitled "On some abnormalities of ocular movements". In this paper he recapitulated his concept of different representations of movements in centres at various levels of the nervous system, this time specifically applied to eye movements. The first level comprised a direct representation of eye movements in the ocular motor nuclei of the brain stem. A second, indirect representation or re-representation of ocular muscles as they are involved in more complex movements was encoded in centres of the middle level, which he interpreted as the motor region of the cerebral cortex. Finally, a third, "doubly indirect" and still more complex representation occurred at the highest level, the prefrontal lobe, containing, as he termed them, the "mental centres". Thus he suggested a hierarchical control system in which motor impulses depart from the prefrontal lobe and reach the ocular motor periphery via the middle and lowest levels.

Jackson was strongly influenced in his views of cortical representation by his clinical observations: patients with frontal lobe damage show conjugate eye deviation to the side of the lesion and away from the hemiplegia, and conversely convulsions due to cortical lesions produce a contralateral conjugate ocular deviation. Also available at the same time was additional evidence obtained from animals, indicating the importance of cortical areas in the frontal lobe for the generation of eye movements. In particular, Ferrier (1874) had observed the effects of electrical stimulation of the frontal eye field (FEF) in monkeys, which resulted in contralaterally directed eye movements.

In the intervening 80 years or so since Jackson's paper a system of neural centres with complex interconnections involved in ocular motor control has been unravelled. Since more is known about the neural control of saccades than any other type of eye movement, this chapter will discuss the major centres involved in their generation – the FEF, superior colliculus, parietal lobe and substantia nigra, pars reticulata – and review the extent to which Jackson's concept of motor hierarchies accords with current views. Whether or not "higher" saccadic centres are considered to be arranged in a hierarchical or parallel fashion, the final common pathway to the ocular musculature is via the brain stem saccadic generator. A brief outline of the structure and function of this saccadic generator is necessary before consideration of these "higher" centres. For horizontal eye saccades to take placc in a yoked conjugate manner it is obviously important to have simultaneous, or near simultaneous, activation of the ipsilateral lateral rectus and contralateral medial rectus muscles with concurrent inhibition of the antagonist muscles. Recent neuroanatomical studies (Büttner-Ennever and Akert 1981) have shown that this is achieved by the presence in the abducens nuclei, not only of motor neurons innervating the ipsilateral lateral rectus muscle, but also interneurons whose axons cross the midline and ascend in the medial longitudinal fasciculus (MLF) to synapse directly on the motor neurons innervating the contralateral medial recti. Thus conjugate ipsilateral horizontal eye movements are facilitated merely by activation of both sets of neurons in a single nucleus (Baker and Highstein 1975). The immediate premotor neurons for these nuclei lie in the paramedian pontine reticular formation (PPRF) which extends from the rostral pole of the abducens nucleus to the pontopeduncular junction lying ventral to the MLF. These premotor or "burst" neurons discharge before and during a saccade but remain silent during fixation and slow eye movements (for review see Fuchs et al. 1985). Single cell recordings show a high-frequency discharge or pulse, the duration of which determines the saccade amplitude. This is following by a step change in the tonic discharge rate which maintains the eye in its new position countering the viscoelastic forces of the orbital contents. The step originates either from the nucleus prepositus hypoglossi, or from the medial vestibular nucleus or simultaneously from these two structures (Cannon and Robinson 1987). A group of neurons called "omnipause" cells located rostrally and medially in the PPRF have been found to discharge tonically, so inhibiting burst neurons at all times, except when an appropriate saccade is required and they cease to discharge (Keller 1974; King et al. 1980). Generation of a saccade by higher centres, therefore, requires inhibition of the omnipause cells and activation of the burst neurons by afferent inputs from higher centres.

Saccadic eye movements are, of course, generated under a number of different circumstances: for example, quick phases of vestibular and optokinetic nystagmus, in foveating a target of interest in the visual field, or in spontaneous saccades in the dark. All these types of saccade use the same promotor programme (Fuchs et al. 1985). The brain stem saccade generator is therefore able to make the necessary calculations of the forces to be exerted by each extraocular muscle, so that the higher centres merely have to signal direction and amplitude to elicit the required saccade.

Turning now to the first of the higher centres, the FEF, it was Ferrier (1874, 1886) who first electrically stimulated the frontal lobe of monkeys, and localised an area which, when stimulated, led to contralateral conjugate eye deviation. These studies led to the concept that head and eye turning during focal seizures probably resulted from an electrical discharge in the contralateral frontal lobe (Holmes 1938). Many subsequent studies of electrical stimulation, both in primates (Wagman et al. 1961; Robinson and Fuchs 1969; Bruce et al. 1985) and in man (Penfield and Jasper 1954) have confirmed Ferrier's original observations, but suggest that the apparent anatomical extent of the frontal eye field depends on the amount of current used. Early studies did not measure current directly, and eye movements were elicited by stimulation over much of the frontal cortex. The FEF appears much smaller and is confined to the arcuate sulcus when small currents are used with stimulating microelectrodes in the grey matter. These electrically elicited eye movements are saccades: they are indistinguishable from the animals's own natural saccades in that they show similar functional relationships between amplitude, duration and velocity. The other important observation from these studies is that the amplitude and direction of the saccade elicited is dependent only upon the specific location of the stimulation within the FEF, and not upon the amount of current used.

These studies suggested a role for the FEF in saccadic initiation; however, when the first single unit recordings were made in primates by Bizzi and associates (1968, 1970), they failed to reveal any neurons which discharged prior to eye movements, though some did discharge after the onset of a saccade. Unfortunately, these studies concentrated on spontaneous saccades occurring in the dark, which is only one condition in which saccades are produced. Recently, Bruce and Goldberg (1985) have found neurons in the FEF which do discharge before saccades, but only under specific conditions in which voluntary saccades are made to actual or remembered visual targets. These presaccade-related neurons usually have visual receptive fields which show no specificity for shape, colour, orientation etc. About 40% of them merely respond transiently when a visual stimulus falls in their receptive field, and are called visual cells. Conversely, 20% discharge before saccades even in the absence of visual stimulation, called movement cells. The remaining 40%, classified as visuomotor cells, show a discharge when the stimulus appears, which persists and enhances just prior to the saccade. No FEF neurons have been found to discharge prior to saccades made spontaneously in the dark or when viewing a simple scene.

In man, support for the role and localisation of the FEF in saccadic generation has been obtained from regional cerebral blood flow studies. These have shown increased blood flow within the FEF, supplementary motor area and the cerebellum during repeated saccades (Melamed and Larsen 1979; Fox et al. 1985). The FEF were identified as discrete cortical regions, consistently active during the generation of voluntary saccades and uninfluenced by target presence, type of cue, or task complexity, this indicating a predominantly motor function (Fox et al. 1985).

If the FEF is involved in saccade initiation it must have appropriate connections with the brain stem saccade generator described earlier. Recent neuroanatomical tracer studies have shown that projections from the FEF course in the anterior limb of the internal capsule and then diverge in the rostral diencephalon into two limbs, one being a transthalamic limb to the thalamic structures, pretectum and superior colliculus. The other is a penduncular limb to the rostral interstitial nucleus of the MLF and the PPRF (Distel and Fries 1982; Leichnetz et al. 1984; Schnyder et al. 1985); these two structures are the immediate premotor areas for vertical and horizontal saccades respectively.

The FEF, therefore, appears to have an important role in the initiation of voluntary saccades, particularly when made to a visual target. Another area shown by cerebral blood flow studies to be active during saccades lies close to the FEF in a part of the supplementary motor area (SMA). Earlier electrophysiological mapping studies of the SMA revealed neurons related to eye movements (Woolsey et al. 1952; Brinkman and Porter 1979), and a readiness potential was noted over the SMA region prior to the onset of saccades (Becker et al. 1972). However, the precise properties of saccade-related neurons have only recently been revealed (Schlag and Schlag-Rey 1987). Electrical stimulation of a small area of dorsomedial frontal cortex in monkeys, called the supplementary eye field (SEF), evoked contraversive saccades, and presaccadic unit activity was recorded with self-initiated as well as visually triggered saccades. Since both the FEF and SEF project to the superior colliculus and the pontine premotor oculomotor area (Leichnitz et al. 1984), their roles in saccade generation may differ. In Jackson's scheme the SEF would be at the highest level; indeed, it appears to be specialised for saccade initiation based on internal cues. It seems unlikely, however, that there is serial processing between the SEF and FEF since neurons in the latter only discharge prior to visually induced saccades; they never discharge in relation to self-initiated saccades as do neurons in the SEF.

The remaining cortical area implicated in the control of saccades in monkeys is an area of the parietal-occipital cortex, the confines of which have not been adequately determined. It is likely to include a portion of cortex in the caudal bank of the intraparietal sulcus that has been called variously POa, VIP and LIP. It receives a projection from visual areas V2 and the medial temporal area (MT) and has a dense projection to the intermediate and deep layers of the superior colliculus (Lynch et al. 1985). In addition, POa contains neurons that discharge prior to visually guided saccades, and stimulation of this region results in contraversive saccades (Shibutani et al. 1984).

Along with these cortical areas the other neural structure most closely associated with the generation of saccades is the superior colliculus (reviewed in Wurtz and Albano 1980), stimulation of which was shown over 100 years ago to produce eye movements (Adamuk 1870). The superficial layers receive a strong visual input both directly from the retina and indirectly from the occipital cortex. The intermediate layers of the superior colliculus receive inputs from the FEF and the SEF. The deep layers project to many of the same midbrain, pontine and thalamic structures as does the FEF. A similarity to the FEF is the finding that electrical stimulation of these deep layers also results in contraversive saccades (Schiller and Stryker 1972).

Single cell recordings in the intermediate and deep layers of the superior colliculus show that neurons mostly discharge before saccades made under any condition (Schiller and Stryker 1972; Mays and Sparks 1980). Each cell, as in the FEF, has a "movement field" so that it discharges only in relation to saccades to a certain area in the visual field but not to any other. This "movement field" is analagous to the "receptive field" of the visual neurons in the superficial layer.

One possibility is that the colliculus acts merely as an intermediate oculomotor centre in a hierarchical system relaying inputs from the FEF and other cortical areas to the brain stem saccade generator. That this cannot be its only role is shown by the continued generation of saccades by electrical stimulation of the FEF following ablation of the colliculi in monkeys. However, after such a lesion stimulation of the posterior parietal or occipital cortex does fail to elicit a saccade (Keating et al. 1983).

Another important centre involved in saccadic control is the basal ganglia. This is suggested clinically by the observation of abnormal saccades in patients with Parkinson's disease (White et al. 1983; Gibson et al. 1987) and Huntington's disease (Leigh et al. 1983). The anatomical connections of the basal ganglia indicate a direct input from the FEF to the body of the caudate nucleus. An outflow from this nucleus via the substantia nigra, pars reticulata (SNpr) projects directly to the superior colliculus and brain stem tegmentum (Jayaraman et al. 1977). Direct evidence for the basal ganglia's role in saccade generation has come from a recent series of experiments in monkeys in which Hikosaka and Wurtz (1983a–d, 1985a,b) have shown that cells in the SNpr exert a tonic inhibition on the saccade-related neurons in the superior colliculus. These nigral cells are GABA-ergic and usually discharge with high frequencies, except just before a saccade made under appropriate conditions. Such conditions include saccades made to visual targets or, interestingly, saccades made to the location of a target briefly presented and then remembered. On the other hand, saccades made spontaneously in light or dark are not accompanied by any modulation of the discharge rate. This decrease in nigral cell activity correlates with an increase in collicular cell activity in the area to which it projects.

The evidence thus far suggests that during steady fixation the SNpr acts as a gating mechanism and maintains an inhibitory effect on the superior colliculus, thereby preventing extraneous involuntary saccades. It is still unclear whether or not nigral cells play a permissive role by reducing their inhibition, so allowing other important influences from the FEF and parietal-occipital cortex to trigger collicular cells. It may equally well act by releasing intrinsic saccade generation mechanisms within the colliculus itself from inhibition. One prediction from this latter hypothesis is that decreased activity or destruction of the SNpr might lead to the generation of inappropriate saccades. Interestingly, patients with certain basal ganglia disorders such as Huntington's disease, which leads to degeneration in the caudate-nigral pathways, do show excessive extraneous saccades when trying to fixate (Leigh et al. 1983).

Although recording of saccade-related neuronal discharges and results of microstimulation indicate a particular centre's involvement in the generation of saccades they do not establish that centre's relative importance during normal behaviour. To help determine this experimenters have turned to the classic approach: ablation.

Hughlings Jackson described how destructive ischaemic lesions of the frontal lobes in man result in conjugate eye deviation towards the side of the lesion. Clinically, this is short lived, usually lasting less than 6 days, after which the saccades appear to be normal on bedside testing. This is in contrast to other accompanying focal neurological signs which may persist for considerably longer periods.

In a recent clincial study Steiner and Melamed (1984) observed that 6 out of 42 patients with acute hemiplegic strokes showed prolonged conjugate eye deviation, and these were the only patients who had had previous lesions affecting their contralateral FEF. It has also been found that patients can make contralateral saccades during intracarotid sodium amytal injections, but that saccades to visual stimuli are performed more easily than those to verbal commands alone (Lesser et al. 1985). These observations have been interpreted as either the intact ipsilateral superior colliculus or the contralateral FEF taking over the function of the affected FEF. Other interesting clinical observations in patients with frontal lobe lesions were made by Holmes (1938), who noted that such patients had difficulty obeying verbal commands such as "eyes right" or "eyes left", and by Luria (1966), who emphasised a lack of sophisticated scanning of complex visual scenes. Finally, Guitton et al. (1985) have recently carried out an interesting study of saccades in patients with discrete removals of frontal lobe tissue for the relief of intractable epilepsy. Using an "antisaccade" task the subjects were required *not* to look at the location of the visual stimulus which appeared in the peripheral field, but in an equal and opposite direction. These patients had considerable difficulties in suppressing disallowed glances directly to the visual stimuli which suddenly appeared in their peripheral vision. It was concluded that the FEF is involved in suppressing unwanted reflex-like oculomotor activity, and also in triggering the appropriate volitional movements when the goal for the movement is known but not yet visible.

Behavioural studies in primates after ablation of the superior colliculus and FEF separately or in sequence have also been performed. Schiller and colleagues (1980) performed experiments in which a Perspex sheet with nine holes, each containing a piece of apple, was moved in front of an animal trained to retrieve each piece. The monkey, with head fixed, would make repeated saccades to each hole before grasping the corresponding piece of apple. After bilateral ablation of the FEF there was initially a pronounced deficit in looking at peripheral targets. Two weeks later the monkey's behaviour returned to normal, and almost normal saccadic scanning of the Perspex sheet was seen. Presumably some other neural structure, such as the superior colliculus, must have taken over the FEF function.

Turning to the effect of bilateral collicular ablation in monkeys, Schiller et al. (1980) found, after 4 postoperative days, a moderate reduction in the number and amplitude of saccades made, which persisted with little recovery. In other experiments, if saccades to distracting peripheral stimuli were not rewarded, such lesioned animals showed an even greater reduction in the number of saccades (Albano et al. 1982). This behavioural deficit can be interpreted as resulting from a superior collicular input, which would normally facilitate either the selection of a visual target for a saccade, or the initiation of a visually guided saccade.

The effect of paired sequential ablation of both the FEF's and superior colliculi in monkeys, however, produced a striking and lasting deficit. These animals were unable to make saccadic eye movements to peripheral targets, and the ocular motor range of their spontaneous eye movements was limited to ±10 degrees (Schiller et al. 1980). This experiment strongly supports the importance of both the FEF and the superior colliculus in the control of saccades and implies that they represent parallel pathways to the brain stem saccadic generator. However, there is another possible interpretation, which is that the deficit merely reflects a non-specific depression of brain stem ocular motor centres after their major motor inputs have been removed (Zee 1983).

From these anatomical, physiological and behavioural experiments it does appear that the superior colliculus and the FEF are intimately involved in visual search behaviour. In this behaviour there is a requirement for the localisation of a visual target in the periphery, and for the eyes to be moved so as to foveate it. This is presumably mediated by the superior colliculi. At a higher level there must be a system responsible for a systematic voluntary internally organised scanning of the enviroment to decide which of several targets, each of equal potential for triggering the foveation mechanism, should be explored next. The FEF and SEF may presumably play a role in this behaviour.

Hughlings Jackson's concept of hierarchies in the light of these observations now appears to be too polarised. Although the control of saccadic eye movements does depend on multiple representations, the lowest being in the brain stem, "higher" level representations are more complex and depend on parallel connections between a variety of neural structures. Single cell electrophysiological recordings along with ablation studies have suggested that the FEF and superior colliculi provide parallel inputs, although amongst the multiplicity of different saccadic tasks in an animal's behavioural repertoire the FEF appears to be more concerned with voluntary saccades, whereas the colliculus is concerned with reflex changes of gaze. A third "higher" centre which may be involved is the

SEF, which is involved with both visually triggered and spontaneous saccades. Finally, the precise role of the SNpr in effecting a gating mechanism for saccades is still uncertain.

References

Adamuk E (1870) Ubere die Innervation der Augenbewegungen. Zentralbl Med Wiss 8:65
Albano JE, Mishkin M, Westbrook LE, Wurtz RM (1982) Visuomotor deficits following ablation of monkey superior colliculus. J Neurophysiol 48:338–350
Baker R, Highstem SM (1975) Physiological identification of interneurons and motorneurons in the abducens nucleus. Brain Res 91:292–298
Becker W, Hoehne O, Iwase K, Kornhuber HH (1972) Bereitshaft potential, pramotorische Positivierung und andere Hirnpotentiate bei sakkadischen Augenbewegungen. Vis Res 12:421–436
Bizzi E (1968) Discharge of frontal eye field neurons during saccadic and following eye movements in unanaesthetised monkeys. Exp Brain Res 6:69–80
Bizzi E, Schiller PH (1970) Neuronal activity in the frontal eye fields of unanaesthetised monkeys during head and eye movement. Exp Brain Res 10:151–158
Brinkman C, Porter R (1979) Supplementary motor area in the monkey: activity of neurons during performance of a learned motor task. J Neurophysiol 42:681–709
Bruce CJ, Goldberg ME (1985) Primate frontal eye fields. I. Single neurons discharging before saccades. J Neurophysiol 53:603–635
Bruce CJ, Goldberg ME, Bushnell MC, Stanton GB (1985) Primate frontal eye fields II. Physiological and anatomical correlates of electrically evoked eye movements. J Neurophysiol 54: 714–734
Büttner-Ennever JA, Akert K (1981) Medial rectus subgroups of the oculomotor nucleus and their abducens internuclear input in the monkey. J Comp Neurol 197:17–27
Cannon SC, Robinson DA (1987) Loss of the neural integrator of the oculomotor system from brain stem lesions in monkey. J Neurophysiol 57:1383–1409
Distel H, Fries W (1982) Contralateral cortical projections to the superior colliculus in the macaque monkey. Exp Brain Res 48:157–162
Ferrier D (1874) The localisation of function in the brain. Proc R Soc Lond (Series B) 22:229–232
Ferrier D (1986) Functions of the brain. Smith, Elder, London
Fox PT, Fox JM, Raichle ME, Burder RM (1985) The role of cerebral cortex in the generation of voluntary saccades: a positron emission tomographic study. J Neurophysiol 54:348–369
Fuchs AF, Kaneko CRS, Scudder CA (1985) Brainstem control of saccadic eye movements. Annu Rev Neurosci 8:307–337
Gibson JM, Pimlott R, Kennard C (1987) Ocular motor and manual tracking in Parkinson's disease and the effect of treatment. J Neurol Neurosurg Psychiatry 50:853–860
Guitton D, Buchtel HA, Douglas RM (1985) Frontal lobe lesions in man cause difficulties in suppressing reflexive glances and in generating goal-directed saccades. Exp Brain Res 58:455–472
Hikosaka O, Wurtz RH (1983a) Visual and oculomotor functions of monkey substantia nigra pars reticulata. I Relation of visual and auditory responses to saccades. J Neurophysiol 49:1230–1253
Hikosaka O, Wurtz RT (1983b) Visual and oculomotor functions of monkey substantia nigra pars reticulata. II Visual responses related to fixation of gaze. J Neurophysiol 49:1254–1267
Hikosaka O, Wurtz RH (1983c) Visual and oculomotor functions of monkey substantia nigra pars reticulata. III Memory contingent visual and saccade responses. J Neurophysiol 49:1269–1284
Hikosaka O, Wurtz RH (1983d) Visual and oculomotor functions of monkey substantia nigra pars reticulata. IV Relation of substantia nigra to superior colliculus. J Neurophysiol 49:1285–1301
Hikosaka, O, Wurtz RH (1985a) Modification of saccadic eye movements by GABA-related substances. I Effect of muscimol and bicuculline in monkey superior colliculus. J Neurophysiol 53:266–291
Hikosaka O, Wurtz RH (1985b) Modification of saccadic eye movements by GABA-related substances. II Effects of muscimol in monkey substantia nigra pars reticulata. J Neurophysiol 53:292–308
Holmes G (1938) The cerebral integration of the ocular movements. Br Med J II:107–112

Jackson JH (1909) On some abnormalities of ocular movements. Lancet I:900–912
Jayaraman A, Batton RR III, Carpenter MB (1977) Nigrotectal projections in the monkey: an autoradiographic study. Brain Res 135:147–152
Keating EG, Gooley SG, Pratt SE, Kelsey JE (1983) Removing the superior colliculus silences eye movements normally evolved from stimulation of the parietal and occipital eye fields. Brain Res 269:145–148
Keller EL (1974) Participation of medial pontine reticular formation in eye movement generation in monkey. J Neurophysiol 37:316–332
King WM, Precht W, Dieringer N (1980) Afferent and efferent connections of cat omnipause neurons. Exp Brain Res 38:395–403
Leichnetz GR, Spencer RF, Smith DJ (1984) Cortical projections to nuclei adjacent to the oculomotor complex in the medial dienmesencephalic tegmentum in the monkey. J Comp Neurol 228:359–387
Leigh RJ, Newman SA, Folstein SE, Lasker AG, Jensen BA (1983) Abnormal ocular motor control in Huntington's disease. Neurology 33:1268–1275
Lesser RP, Leigh RJ, Dinner DS, Luden H, Morris HH, Tomsak RL, Lockwood KI (1985) Preservation of voluntary saccades after intracarotid injection of barbiturates. Neurology 35:1108–1112
Luria AR (1966) Higher cortical functions in man. Translated by Haigh B. Basic Books, New York, pp 293–327
Lynch JC, Graybiel AM, Lobeck LJ (1985) The differential projection of two cytoarchitectonic subregions of the inferior parietal lobule of macaque upon the deep layers of the superior colliculus. J Comp Neurol 235:241–254
Mays LE, Sparks DL (1980) Dissociation of visual and saccade-related responses in superior colliculus neurons. J Neurophysiol 43:207–231
Melamed E, Larson B (1979) Cortical activation pattern during saccadic eye movements in humans: localisation by focal cerebral blood flow increases. Ann Neurol 5:79–88
Penfield W, Jasper H (1954) Epilepsy and the functional anatomy of the human brain. Little, Brown, Boston
Robinson DA, Fuchs AF (1969) Eye movements evoked by stimulation of frontal eye fields. J Neurophysiol 32:637–648
Schiller PH, Stryker M (1972) Single-unit recording and stimulation in superior colliculus of the alert rhesus monkey. J Neurophysiol 35:915–924
Schiller PH, True SD, Conway JL (1980) Deficits in eye movements following frontal eye-field and superior colliculus ablations. J Neurophysiol 44:1175–1189
Schlag J, Schalg-Rey M (1987) Evidence for a supplementary eye field. J Neurophysiol 57:179–200
Schnyder H, Reisine H, Hepp K, Henn V (1985) Frontal eye field projection to the paramedian pontine reticular formation traced with wheat germ agglutinin in the monkey. Brain Res 329:151–160
Shibutani H, Sakata H, Hyvarinen J (1984) Saccade and blinking evoked by microstimulation of the posterior parietal association cortex of the monkey. Exp Brain Res 55:1–8
Steiner I, Melamed E (1984) Conjugate eye deviation after acute hemispheric stroke: delayed recovery after previous contralateral frontal lobe damage. Ann Neurol 16:509–511
Wagman IH, Krieger HP, Papatheodorou CA, Bender MB (1961) Eye movements elicited by surface and depth stimulation of the frontal lobe of Macaque mulatta. J Comp Neurol 117:179–188
White OB, Saint-Cyr JA, Tomlinson RD, Sharpe JA (1983) Ocular motor deficits in Parkinson's disease. II. Control of the saccadic and smooth pursuit systems. Brain 106:571–587
Woolsey CN, Settlage PH, Meyer DR, Spencer W, Hamuy TP, Travis AM (1952) Patterns of location in the precentral and 'supplementary' motor area and their relation to the concept of a premotor area. Res Publ Assoc Res Nerv Ment Dis 30:238–264
Wurtz RH, Albano JE (1980) Visual-motor functions of the primate superior colliculus. Annu Rev Neurosci 3:189–226
Zee DS (1984) Ocular motor control: the cerebral control of saccadic eye movements. In: Lessell S, van Dalen JTW (eds) Neuropathalmology 1984, vol 3, Elsevier, Amsterdam, pp 141–156

Chapter 18

Hierarchies in the Cerebellum

J. F. Stein

I hope to show that the cerebellum is extremely important in transforming visual information in order to control motor output. Clearly this is as relevant to ocular motor control described in chapter 17, as it is to limb movements.

Hughlings Jackson never really came to terms with the cerebellum (Taylor 1931/32). I believe it is true to say that he had no clear idea of how the cerebellum fitted into his hierarchical scheme of how the nervous system is functionally organised. What I propose to attempt in this chapter, therefore, is to show that actually the cerebellum does have an important part to play in Hughlings Jackson's scheme of things, mediating important transfers of information between the levels that he described.

There have been concerted attacks on Jackson's concept of hierarchical levels in the nervous system, emanating particularly from sensory physiologists. But in fact even they use his terminology. Likewise, motor physiologists now recognise that the brain is a democracy rather than a dictatorship, in which control of the interacting loops which are thought to guide movement passes both upwards and downwards, rather than solely centrifugally. Nevertheless, they still employ fundamentally the same ideas that Hughlings Jackson introduced.

I shall take as an example of motor control, the visual guidance of movement. I shall show that the visual information necessary for guiding limb movements reaches motor centres not by purely intracortical, cortico-cortical, connections but mainly via the cerebellum (Gibson et al. 1980).

It is often assumed that there are important connections linking visual areas in the occipital cortex directly with the motor cortex, and that these underlie the visual control of limb movements. In reality, however, there are no direct connections between visual and motor cortices. There are the indirect projections from parieto-occipital cortex to the frontal eye fields that have already been discussed (see Chap. 17, p. 153), and from there run connections to the premotor cortex and thence to motor cortex proper (Pandya and Kuypers 1969). But there are no direct connections between visual and motor cortices.

Confirming these anatomical results behaviourly, Myers et al. (1962) showed some time ago that cortico-cortical connections are not essential for the visual guidance of limb movements. In monkeys they divided the superior longitudinal fasciculus, thus cutting all the axons that pass between the parieto-occipital cortex behind the central sulcus and the motor areas in front. They made a complete cortico-cortical disconnection, without destroying subcortically directed fibres. Nevertheless the monkeys' ability to guide their limbs visually was hardly affected. They were able to pick up small pieces of apple even from a moving turntable; the monkeys were not impeded in any obvious way in any of the visuomotor tests that Myers and Sperry devised. These experiments strongly suggested that structures beneath the cortex are more important than cortico-cortical connections for linking vision with movement. Of course the most significant of these subcortical motor structures is the cerebellum, but in passing I must draw attention to another potentially important subcortical pathway, that involving the basal ganglia (Kemp and Powell 1970; Wise and Evarts 1985). These demand more than the brief sentences which I can devote to them here, and they are well covered in Edmund Rolls' chapter (see Chap. 15). So I shall not discuss them, beyond asserting that probably the basal ganglia are not involved in the sensory guidance of movement, the continuous visual monitoring of action; rather they may be concerned with the behavioural triggering of complete pre-programmed motor acts (Stein 1985).

My task is to describe how the cerebellum mediates between Hughlings Jackson's levels of control. The visual association areas in parieto-occipital cortex form part of what Hughlings Jackson termed the "highest" level of motor control, the sensorimotor association areas. By tracing the outputs of these areas to motor structures we can begin to answer the very important question; where do primary visual processing areas project onwards (Glickstein 1972)? I shall concentrate on the efferent connections of movement-sensitive visual areas in prestriate (V2 and V5) and posterior parietal cortex (area 7). The main outputs of these areas are very dense projections to the pontine nuclei (Brodal 1972; Glickstein et al. 1972; Baker et al. 1976), and thence to the cerebellum (Mower et al. 1982; Fig. 18.1).

The cerebellum can be divided longitudinally into three main functional regions, a lateral cortical area, whose output is relayed via the dentate nucleus; an intermediate area, the paravermal cortex, whose main output nucleus is the interpositus (globus and emboliform in humans); and the vermal cortex, including the flocculonodular lobe, whose main output is to the fastigial nucleus together with the lateral vestibular (Deiter's) nucleus. There are approximately 21 million fibres leaving the cerebral cortex which pass into the midbrain via the cerebral peduncles. Of these 21 million axons, perhaps 20 million go no further than the pontine nuclei, relaying solely to the cerebellar cortex. In addition, most of the million axons which do pass caudally into the medullary pyramids and then into the corticospinal tract give collaterals to the cerebellum (Cajal 1909). So this is not a minor pathway, but quantitatively one of the most important causeways in the whole brain.

Visual association areas V2, V5 and area 7 connect mainly with the lateral cerebellar cortex (Mower et al. 1982; Fig. 18.1). The Purkinje cells here inhibit dentate neurons onto which their axons project (Eccles et al. 1967). Dentate neurons in turn project onwards via the lateral thalamus to the motor cortex. Thus the lateral cerebellum forms a potentially very important link passing visual information from parieto-occipital visual association areas to the motor cortex. The reader will note that these connections enable the transfer of visual information from Hughlings Jackson's highest level of the motor hierarchy, the sensorimotor association areas, to his next lower – middle – level, the motor cortex.

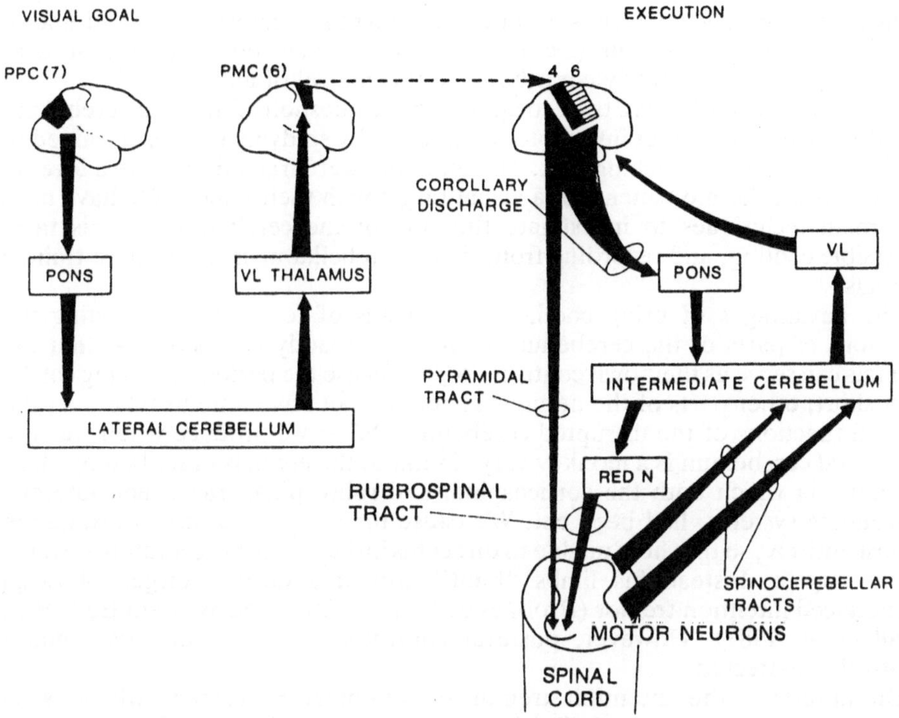

Fig. 18.1. Visuocerebellar and corticomotoneuronal connections mediating the visual guidance of voluntary movements.

The motor cortex in turn projects via the pontine nuclei back to the paravermal region of the cerebellum (Brodal 1972; Fig. 18.1). This region also receives the spinocerebellar tracts, and it projects via the interpositus nucleus to the red nucleus or, via the ventrolateral nucleus of the thalamus, to the motor cortex again. Thus the lateral cerebellum relays visual signals to the motor cortex. Thence the corticospinal tract gives the motor cortex potentially direct access to the spinal cord, or the motor cortex can address motoneurones indirectly via its connections with intermediate cerebellum and the red nucleus and rubrospinal tract (Fig. 18.1). The corticospinal and rubrospinal tracts together comprise Hans Kuypers' lateral brain stem descending system for controlling distal movements of the limbs (see Chap. 14, p. 113).

The vermal and flocculonodular parts of the cerebellum connect the vestibular system with Kuypers' medially descending pathways for activating axial muscles, regulating posture via the reticulospinal and vestibulospinal tracts. The motor neurons and reflex machinery in the brain stem and spinal cord constitute Hughlings Jackson's lowest level.

Thus recent anatomical findings (Fig. 18.1) suggest how Hughlings Jackson might have fitted the cerebellum into his motor hierarchy. Unfortunately, however, they tell us very little about what the cerebellum actually does. Regrettably, despite the huge amount that we now know about the connections of the cerebellum (Ito 1984), and also the very great advances, which we owe to Sir

John Eccles and his colleagues, in understanding the microcircuitry and microphysiology of the cerebellum (Eccles 1967), we still have only the vaguest ideas how the cerebellum actually contributes to the control of movement.

My coworkers and I have tried to approach the question of how the cerebellum transforms its visual input into motor commands by studying monkeys trained to perform a visumotor tracking task. The monkeys were trained to move a lever in order to track the movements of a visual target with their hands. We have used two main techniques to investigate the role of the cerebellum in this task: reversible cooling, and recording from single cerebellar neurons in these trained animals.

The advantages of using cooling as a means of temporarily arresting the functions of parts of the cerebellum is that it is readily reversible, so that the animals can serve as their own controls. Also because the period of cooling can be kept short, other parts of the nervous system do not have time to take over the normal functions of the disrupted cerebellum. What we produce when we cool the lateral cerebellum is a monkey very similar to the animals described by Hans Kuypers, in which both the corticospinal and rubrospinal tracts, and laterally descending systems, had been cut. We cause the monkey to lose coordination almost entirely, so that he is unable to direct his ipsilateral arm accurately towards visual targets. Instead his limbs "hunt" around a desired target, showing pronounced intention tremor (Brookes et al. 1973; Stein and Wattam Bell 1975; Miall et al. 1987). However, postural control of axial movements remains relatively unaffected.

On observing the intention tremor of voluntary movement which is so characteristic of lateral cerebellar lesions, Holmes (1939) speculated that the cerebellum may be concerned in the feedback control of movement. He suggested that the cerebellum may be a "comparator" comparing the desired position of a limb, supplied by the motor cortex, with its actual position, fed back via the spinocerebellar tracts. Lesions of the cerebellum cause intention tremor, it was suggested, because accurate calculation of the mismatch between arm and target is prevented. Hunting is not a sign that a feedback system has been completely destroyed, rather that it has departed from perfect adjustment. Negative feedback and a comparator process are still operating. What intention tremor suggests is that cerebellar lesions prevent proper adjustment of the operating parameters of the control system from taking place.

When considering these problems, one must remember that there is an irreducible delay between visual input and motor output, not measured in terms of a few milliseconds, but somewhere around a quarter of a second (Glickstein 1972). However well an animal or a human is trained to make a response to a visual stimulus, there is an irreducible reaction time of some 250 ms. This long time is potentially disastrous for a negative feedback control system, and it is something that the neural network responsible for visuomotor control has somehow to overcome. The cerebellum is probably concerned in at least three ways in helping to avoid these problems.

In order to understand them more clearly I shall first outline the effects of the long-time visuomotor delay in a little more detail. If one tried to design a linear feedback system with a 250 ms delay, together with perhaps 100 ms motor delay, it would not be possible to make it generate accurate movements which were shorter than 3 or 4 s in duration. It is simply impossible to get a negative feedback system to do better. If one tried increasing the gain of the system in order to overcome the loop delays, it would rapidly become unstable, and start oscillating uncontrollably. Thus, if an engineer is faced with an irreducible delay of 350 ms in a feedback system, he has little option but to limit it to making only very slow movements. However, neurological control systems appear to have evolved some very clever tricks, not well known to engineers, to overcome these problems.

The first of these is to employ the strategy of "sampling" (Stark 1968). When a monkey is tracking a smoothly moving target, instead of trying to follow the target continuously he is usually observed to make discontinuous, intermittent movements (Miall et al. 1986). The visual situation is sampled only every 500 ms or so. Humans show the same odd behaviour when tracking unpredictable, continuously moving targets. The advantage of this strategy is that it enables the limb guidance control system to be furnished with the high gain which is required in order to make fast movements, without risking instability. The system does not start to oscillate because cumulative small errors cannot build up. Since the flow of information around the control loop is periodically interrupted by the sampling process, the error state currently existing can be completely corrected.

But merely correcting the current error would not suffice, since more errors would build up during the time that the correction was taking place. So a second trick that neurological control systems have learnt is to introduce prediction. By assessing the current speed of a target, monkeys can predict where it is going to be at any point in the future, e.g. 350 ms later (Miall et al. 1986). Thus they can allow for the probable movement of the target during the delay which the subject knows will take place before his next movement is completed.

We now have evidence that the cerebellum makes important contributions to both these processes. When we cool the lateral cerebellum in monkeys tracking a continuously moving target we find that their movements become inaccurate, mainly because they can no longer make use of information about the speed with which the target is moving to calculate accurately the amplitude required of each movement. Normally its size is chosen not only to eliminate current error but also to allow for movement of the target, which will continue while the computation is taking place and the movement is being carried out. Future motion of the target is predicted from its current velocity and direction and used to calculate the size of the movement which will be required to catch it up. The results of our cerebellar cooling experiments thus suggest that the lateral cerebellum plays a crucial part in these computations (Miall et al. 1987).

When we recorded from the lateral cerebellum in awake monkeys tracking a moving visual target we obtained more evidence in favour of these ideas. Neurons there receive visual inputs; but the visual input is discontinuous (Horvat and Stein 1985). It is probably gated out every 500 ms or so in phase with each intermittent movement that the monkey makes. Thus cerebellar recordings show evidence of a sampling process at work. It is possible, and at this stage I am only speculating, that this gating may be controlled by climbing fibre activity. Granit and Phillips (1956) christened the action of climbing fibres the "inactivation response". One effect of the climbing fibre may simply be to wipe Purkinje cell registers clear of pre-existing simple spike activity in order to prepare the slate for another calculation of the parameters required of the next movement (Gellman et al. 1985). Hence climbing fibre activity may introduce the property of sampling the visual input into the system, and thus help to gain the control advantages discussed earlier.

The second transformation that one can see taking place when one records from visually responsive cells in the lateral cerebellum, is that, as expected, they are particularly sensitive, not to the static position of the target at any moment, but to its motion, its changes in position, or velocity. That is what one would expect of the motion inputs to these cells since they come from posterior parietal and prestriate cortical areas which are sensitive to visual movement. Likewise, the deleterious effects of cooling the lateral cerebellum on predictive control suggest that these visual motion inputs are particularly important. Thus lateral cerebellar neurons seem to code the rate and direction of target motion, rather than its position. This velocity data is what is required for predicting where target and limb are going to be some time in the future. Pellionisz and Llinas (1979) have produced a complicated mathematical model which also attributes to the cerebellum this property of prediction. Whilst not everybody is agreed that the details of their model are appropriate to the cerebellum, I think few would now deny that the cerebellum probably has an important part to play in the process of using higher derivatives of a signal in order to predict what parameters for the next movement are required to catch the target.

In the paravermal part of the cerebellum we have recorded from neurons that receive proprioceptive inputs directly from the spinal cord. These cells are probably more concerned with controlling the actual execution of limb movements than with defining their visual goal. They discharge shortly before movement rather than in relation to visual events, and they are closely correlated with the parameters of each movement that the monkey makes. We have found very high correlations (up to 0.95) between their discharges and the velocity of arm movements in a particular direction (Stein 1978).

Thus, as expected from their connections, the lateral and intermediate regions of the cerebellum are rather different. The lateral cortex receives visual inputs and is concerned with the visual goal and visual guidance of a movement. On the other hand, the intermediate part of the cerebellum is probably more involved with the details of its execution. Hence cells here are more tightly tied to movement parameters and under certain circumstances they receive powerful proprioceptive feedback (Stein 1978; Thach 1968).

Having introduced the ideas of predictive control and parameter adjustment I must briefly allude to the exciting topic of what is sometimes, misleadingly, called motor learning: the problem of whether the cerebellum has some part to play in learning motor skills (Marr 1969). This is the third contribution of the cerebellum to motor control alluded to earlier, and a third trick that the neurological control systems responsible for visuomotor guidance have evolved to overcome the constraints set by slow visual processing and relatively slow movements.

The cerebellum probably has no part to play in learning per se, in the sense that most of us mean by that word. An animal with no cerebellum at all still remembers what sort of actions it needs to make to walk, or what sort of movements it needs to perform to open a box. But it cannot make the required movements very skilfully; it is extremely incoordinated. What is becoming clear is that the cerebellum probably has an important part to play in acquiring and honing motor skills, by optimising the parameters of movement control systems in the light of their past successes or failures, in order to adapt them to the current needs of the animal.

There is a popular model that has been studied in great detail in this context: the vestibulo-ocular reflex (VOR). This is the reflex that keeps the eyes fixed in space whenever one moves one's head, by moving them through exactly the same angle as the head, but in the opposite direction. Gonshor and Melville Jones (1975) showed that if one wears reversing spectacles, one can actually reverse the direction in which this reflex operates, so as to make it visually useful once more. If one wears reversing spectacles the normal VOR becomes maladaptive. The reversing spectacles make the world appear to move in the same direction as the head, but the normal VOR moves the eyes on the opposite direction. However, after wearing the spectacles for a few hours the reflex adapts, so now the eyes move in the same direction as the head, rather than opposite. One way of describing this is to say that the VOR changes its gain under these circumstances from -1 to $+1$.

To cut a long story short, it is now fairly clear that part of the cerebellum, the flocculonodular lobe, is very intimately concerned with this adapting process. If this region is lesioned (Takemori and Cohen 1972), plasticity of the VOR disappears, and adaptation no longer takes place. Ito and his colleagues have made recordings from flocculonodular cells in rabbits while adaptation of the VOR is taking place (Ito 1984). He reports that the discharge of these neurons changes in the way predicted by the hypothesis that the cerebellum actually effects and sustains the long-term changes in the parameters of the VOR which are observed. Others are not so sure that the actual plastic change takes place in the cerebellum (Miles et al. 1980; Llinas and Simpson 1981). But many people now believe that the cerebellum probably plays a vital, but not necessarily exclusive, role in such long-term changes (Gilbert and Thach 1977; Hore and Vilis 1984; Ito 1984; McCormack and Thompson 1984; Yeo et al. 1985).

It is tempting to generalise from these ideas about long-term optimisation of the VOR by the cerebellum to suggest that the cerebellum may be intimately concerned with the optimisation of all neurological motor control systems. Clearly a self-optimising adjustor of the parameters of every control loop (metacontroller) would end up improving every motor skill. Supporting these speculations there is suggestive evidence implicating the cerebellum in the metacontrol of systems as different from each other as conditioning of the nictitating membrane reflex (Yeo et al. 1985), postural reactions (Nashner 1976), baroreceptor responses (Moruzzi 1950) and the stretch reflex of skeletal muscles (Mackay and Murphy 1979).

Summary

The cerebellum is a sensory structure mediating between the highest, middle and lowest levels of Hughlings Jackson's hierarchical motor scheme. It is responsible for the sensory guidance of movement, passing information between sensorimotor association and motor areas of the cerebral cortex, and executive motor structures in the brain stem and spinal cord. I have used the example of the visual guidance of movement to demonstrate that functionally the cerebellum probably provides three crucial, but unusual, elements for motor control and the acquisition of motor skills, namely sampling, prediction and optimisation of system parameters.

References

Baker J, Gibson A, Glickstein M, Stein JF (1976) Visual cells in the pontine nucleus of the cat. J Physiol (Lond) 252:415–433

Brodal P (1972) The corticopontine projection from visual cortex in the cat. Brain Res 39:297–335

Brookes VB, Kovsloskaya IB, Atkin A, Horvath FE, Uino M (1973) Effect of cooling dentate nucleus on tracking performance in monkeys. J Neurophysiol 36:974–995

Cajal R (1909) Histologie du système nerveux. C.S.I.C., Madrid

Eccles JC, Ito M, Szentagothai J (1967) The cerebellum as a neuronal machine. Springer, Berlin Heidelberg New York

Gellman R, Gibson AR, Houk JC (1985) Inferior olive neurones in awake cat. Detection of intact and passive body displacement. J Neurophysiol 54:40–60

Gibson A, Mower G, Stein JS (1980) The projection of visual cells in the pons to the cerebellum. J Neurophysiol 43:355–367

Gilbert PFC, Thach WT (1977) Purkinje cell activity during motor learning. Brain Res 128:309–328

Glickstein M (1972) Brain mechanisms in reaction time. Brain Res 40:33–37

Glickstein M, King R, Stein J (1972) Visual input to the pontine nuclei. Science 178:1110–1111

Gonshor A, Melville Jones G (1976) Extreme vestibulo ocular adaptation induced by prolonged optical reversal of vision. J Physiol (Lond) 256:381–414

Granit R, Phillips CG (1956) Excitatory and inhibitory processes on Purkinje cells in the cerebellum. J Physiol 133:520–547

Holmes G (1939) The cerebellum of man. Brain 62:1–30

Hore J, Vilis T (1984) Loss of set in muscle responses to limb perturbations during cerebellar dysfunction. J Neurophysiol 51:1137–1148

Horvat DM, Stein JF (1985) Role of different cerebellar regions in visuomotor control. Neurosci Lett 321:11

Ito M (1984) The cerebellum and neural control. Raven, New York

Kemp JM, Powell TPS (1970) The connexions of the striatum and globus pallidus: synthesis and speculation. Philos Trans R Soc [Biol] 262:441–457

Llinas RR, Simpson JI (1981) Cerebellar control of movement. In: Towe AL, Luschei FS (eds) Handbook of behavioral neurobiology, vol 5. Motor coordination. Plenum, New York, pp 231–302

Marr D (1969) A theory of cerebellar cortex. J Physiol (Lond) 202:437–470

Mackay WA, Murphy JT (1979) Cerebellar modulation of reflex gain. Prog Neurobiol 13:361–417

McCormack DA, Thompson RF (1984) Cerebellum – essential involvement in the classically conditioned eyelid response. Science 223:296–299

Miall RC, Weir DJ, Stein JF (1985) Visuomotor tracking with delayed visual feedback. Neuroscience 16:511–520

Miall RC, Weir, DJ, Stein JF (1986) Manual tracking of visual targets by trained monkeys. Behav Brain Res 20:185–201

Miall RC, Weir DJ, Stein JF (1987) Visuomotor tracking during reversible inactivation of the cerebellum. Exp Brain Res 65:455–464

Miles FA, Fuller JH, Braitman DJ, Dow BM (1980) Long term adaptive changes in primate vestibulocular reflex. III Electrophysiological observations in flocculus of normal monkey. J Neurophysiol 43:1437–1476

Moruzzi G (1950) Problems in cerebellar physiology. Thomas, Springfield, Illinois

Mower G, Gibson A, Robinson F, Stein JF, Glickstein M (1982) Visual ponts cerebellar projections in the cat. J Neurophysiol 43:355

Myers RE, Sperry RW, McCurdy NM (1962) Neural mechanisms in visual guidance of limb movement. Arch Neurol 7:195–202

Nashner LM (1976) Adapting reflexes controlling human posture. Exp Brain Res 26:59–72

Pandya DN, Kuypers H (1969) Cortico-cortical connexions in the rhesus monkey. Brain Res 13:13–36

Pellionisz A, Llinas R (1979) Brain modelling by tensor network theory. The cerebellum. Neuroscience 4:323–348

Stark L (1968) Neurological control systems. Plenum, New York

Stein JF (1978) Long loop motor control in monkeys. In: Desmedt J (ed) Cerebral motor control in man: long loop mechanisms. Karger, Basel, pp 107–122

Stein JF (1985) The control of movement. In: Coen C (ed) Functions of the brain. Oxford University Press, Oxford, pp 67–97

Stein JF, Wattam Bell J (1975) The effect of cooling n. interpositus in rhesus monkeys on the tracking of a visual target. J Physiol (Lond) 252:47P
Takemori S, Cohen B (1972) Visual suppression of vestibular nystagamus. Brain Res 72:203–224
Taylor J (ed) (1931/32) Selected writings of John Hughlings Jackson. Hodder and Stoughton, London. Reprinted (1958) Basic Books, New York
Thach WT (1968) Discharge of cerebellar purkinje and nuclear cells during rapidly alternating arm movements in the monkey. J Neurophysiol 31:785–797
Wise SP, Evarts EV (1985) The motor system in neurobiology. Elsevier, Oxford
Yeo CH, Hardiman MJ, Glickstein M (1985) Classical conditioning of the nictitating membrane response of the rabbit – lesions of the cerebellar cortex. Exp Brain Res 60:99–113

Chapter 19

Sphincter Control Systems in Man

M. Swash

Control of the pelvic sphincters is an essential aspect of the social requirements of everyday life, not only in the human but in other species. Sphincter training is one of the most definite of the recognised milestones of development in infancy. However, as adults we are rarely conscious of the complexities of the normal state of continued continence, and the ability to micturate and defaecate at will. Nonetheless, impairment of these functions is a source not only of embarrassment to the affected individual but also of ostracism and isolation within society. Incontinence thus leads to much misery and distress. The coordinated activity of afferent and efferent pathways in the somatic and autonomic pathways to the sphincter musculature, usually regarded as separately organised and controlled systems, is precise and modulated; yet, until recently, it has received little attention.

The autonomic nervous system is usually thought of as a subconscious control system adapted to the modulation of visceral, vascular, sexual and heat-adaptation responses. Although these mechanisms do not usually reach consciousness, they may do so in certain circumstances, and they may be modified by the actions of the central nervous system. Indeed, recent work suggests a far greater degree of central modulation of these autonomic functions than was previously accepted (Appenzeller and Atkinson 1985; Gillis et al. 1987). The interactions of the brain with the autonomic nervous system extend to sphincter mechanisms, as is well known in the desire to micturate before potentially stressful experiences, such as examinations or athletic events. Studies of the continence mechanisms of the bowel and bladder in the human have tended to neglect this important aspect of the integration of nervous activity and have concentrated, instead, on the autonomic nervous system. Consequently, theories of continence, and of the mechanisms of normal micturition and defaecation have virtually excluded any role of the somatic nervous system. That this must be incorrect is common experience during the acts of micturition and defaecation.

Continence

Urinary continence is maintained by the angle sphincter mechanisms of the bladder neck, which allow the smooth muscle sphincter to operate in preventing the efflux of urine into the proximal urethra. The latter, and bladder distension, are powerful stimuli for initiating the micturition reflex. The angle of the bladder neck is maintained by the action of the anterior pelvic floor muscles. The periurethral striated sphincter muscles are probably particularly important in maintaining these anatomical relationships. Closure of the urethra is additionally maintained by tonic contraction of urethral smooth muscle and by the action of the intramural component of the urethral striated sphincter muscle (Gosling 1979). There is an analogy between the anorectum and urethra in this respect in that the innervation of the periurethral striated sphincter muscle is derived from the perineal branch of the pudendal nerve, and that of the intramural component of the striated sphincter is derived from direct pelvic branches of the sacral plexus; thus the latter innervation resembles that of the puborectalis muscle (Snooks and Swash 1986).

Urinary continence is almost certainly not maintained by a "squeeze" action of the urinary sphincter musculature, since it would require enormous forces to squeeze tightly shut a narrow tube containing urine in the process of expulsion under pressure, but is accomplished by a "kinking" action of these muscles on the urethra. The latter action requires minimal force and would be very effective in arresting or preventing the flow of urine. Indeed, this is the action taken by the domestic gardener in arresting the flow of water through a hosepipe when he is at a distance from the tap.

In the anorectum, faecal continence is maintained by a similar mechanism that depends on the actions of the external anal sphincter muscle and of the puborectalis muscle sling. The latter maintains the angulation of the rectum with the anal canal, forming a flap valve mechanism, so that the anterior and posterior mucosal walls of the anorectum are maintained in close apposition (Parks 1975). Any increase in intra-abdominal pressure as, for example, during a cough, will then apply more tightly the walls of the anorectum, and so continence is maintained.

Micturition and Defaecation

Micturition is accomplished by contraction of the detrusor muscle of the bladder wall, mediated by parasympathetic nervous activity, simultaneous with relaxation of the smooth muscle sphincter of the bladder neck and of the periurethral and intramural striated sphincter musculature. Micturition thus implies the coordinated actions of the autonomic and somatic nervous systems. The stimulus for this micturition activity may be reflex, from stretch afferents in the bladder wall, or may come from gravitational stimuli associated with response originating in the bladder trigone region on assuming the erect posture, or it may be initiated by a volitional desire to micturate of conscious or subconscious derivation, associated, for example, with fear, anxiety or fright.

During defaecation the puborectalis muscle relaxes, as can be shown by EMG studies, and during X-ray proctography, allowing faeces to enter the anal canal. The anal canal contains sensory receptors, and has a low threshold for the discrimination of faecal matter and gas. This sensory input effect reinforces the defaecation reflex, and the complex, coordinated actions of the pelvic floor muscles and of abdominal contractions commences, allowing faecal matter to pass through the anorectum. Both the internal and the external anal sphincter muscles guarding the exit from the anal canal must relax during defaecation (Henry and Swash 1985), a process that can be observed by the curious in any domestic mammalian species. Defaecation may be induced by gastric filling (the so-called gastro-colic reflex), or by colonic activity causing rectal filling. The latter induces the recto-anal reflex, causing anal dilatation, and so changing the subject's perception of the anal canal. The importance of circadian rhythms in patterns of defaecation activity are well known, especially to the jet-lagged traveller awoken by the call to stool at an astonishing local time in a new enviroment. Bowel evacuation may also follow the deliberate, conscious decision to defaecate, thus representing a volitional act, or may follow emotional stimuli.

Although the puborectalis, and the external anal and periurethral striated sphincter muscles are in a state of tonic contraction (Floyd and Walls 1953), mediated reflexly, as discussed above, this can be modulated. For example, a sudden cough causes a transient rise in intra-anal pressure in the sphincter zone by as much as 200 mm water. This increase in intra-anal pressure is due to a brisk increase in contraction of the external anal sphincter, induced reflexly by the cough. The puborectalis and external urinary sphincter muscles respond in a similar fashion to this stimulus. Changes in posture, talking and other activities also cause rapid adjustments in tone in these muscles, designed to maintain apposition of the anal canal and of the urethra, and to maintain the anorectal (Taverner and Smiddy 1959; Parks et al. 1962) and vesical angles. On the other hand, sudden distention of the rectum results in marked inhibition of the contraction of the puborectalis and external anal sphincter muscles, and also of the internal anal sphincter muscle. These are the two (somatic and non-somatic) components of the recto-anal reflex described by Gowers (1877). The smooth muscle component of this reflex involves the enteric nervous system through activation of the myenteric plexus, and does not require synaptic modulation in the spinal cord (Lubowski et al. 1987). After defaecation there is a burst of activity in the external anal sphincter, called the closing reflex, which is accompanied by restitution of the anorectal angle by elevation of the pelvic floor to its former position. These patterns of activation of the pelvic floor and striated sphincter muscles can be studied by EMG (see Swash and Snooks 1985) and by defaecating proctography (Bartram and Mahieu 1985; Womack et al. 1987). Similarly, EMG studies of the urinary striated sphincter musculature with video cystometrograms are used in micturition disorders (Hald and Bradley 1982).

The neural processes subserving these actions of the smooth and striated muscles of the perineum and of the pelvic floor must involve the motoneuronal pools in the sacral cord. They form an interesting and poorly understood interaction of the apparently separate somatic and automatic nervous systems. The many ways in which micturation and defaecation may be initiated suggest that these activities are not controlled from any single part of the brain, but that they may be modulated by stimuli relayed to many different anatomical and functional levels of the central nervous system.

Hierarchies in Sphincter Control Systems

A Jacksonian view of the bladder and bowel implies recognition of levels of control superimposed upon one other, from the lowest to the highest. On this view it is possible to recognise the autonomic and somatic innervations of the smooth and striated components of the sphincter musculature, and of the detrusor mechanisms of the bladder and anorectum, with superimposed levels of function (control mechanisms) overlying this executive system in the spinal cord, brain stem, basal ganglia and cerebral cortex.

An alternative approach to understanding these control systems specifies that they are organised in a network of distributive systems, such that each component can be related to functional motor programs that are interconnected in parallel rather than in series (Ito 1986). Such an arrangement allows the maximal flexibility and capacity for learning and so for change and responsiveness in behaviour since it accepts input for decision-making at several anatomical and functional levels.

Onuf's Sacral Nucleus

The peripheral components of the innervations of these sphincters are organised in the sacral nuclei at the S2 and S3 levels. In this region the motoneurons for the autonomic and somatic efferent nervous systems are situated in close relation to each other, medially in the anterior horn. The specialised groups of somatic efferent anterior horn cells in the ventromedial part of the sacral anterior horn cell pool, forming the Onuf nucleus (Onuf 1900), innervate the striated component of the vesical and anorectal sphincter musculature (Schroder 1980; McKenna and Nadelhaft 1984); they also innervate the anterior perineal muscles, such as the ischiocavernosus and bulbocavernosus muscles (Roppolo et al. 1985). The dorsal component of this nucleus contains motor neurons that innervate the external anal sphincter and the ventral part neurons that innervate the periurethral striated sphincter muscle. The site of the neurons that innervate the puborectalis muscle is unknown, but is probably adjacent to the Onuf nucleus.

The neurons of Onuf's nucleus are smaller than other somatic neurons and have dense dendritic bundles that project rostrocaudally, while remaining within the confines of the column of the nucleus (Roney et al. 1979). These prominent and unusual dendritic arrays provide direct pathways for the interconnection of the neurons of the sacral nucleus. These profuse connections may provide the anatomical basis for the synchronisation of firing of neuronal activity in the nucleus that is so characteristic of its function, for the maintenance of rhythmic and repetitive output and, perhaps, for metabolic and developmental functions (Roney et al. 1979; Roppolo et al. 1985). Other dendrites extend radially from the main body of the nucleus; these make contact with fibres descending rostrocaudally from brain stem and cerebral centres and with afferents from the pudendal nerve, for example muscle spindle and other muscular afferents, and input from the pudendal sensory innervation of the anal canal (Halstege and

Kuypers 1982). Muscle spindles are present, but only in small number in the external anal sphincter muscle (Winckler 1958; Swash 1985), so that primary and secondary afferent input to this nucleus is probably relatively sparse. The neurotransmitters in these terminals are not yet characterised.

Terminals containing leucine-enkaphalin (LEU-ENK), somatostatin (SST) and vasopressin intestinal peptide (VIP) are found in relation to these neurons, but not in the somatic motor neurons in the same sacral segments, just adjacent to the Onuf nucleus itself (Schroder 1984). Similar peptidergic terminals are found in relation to the neurons of the parasympathetic neurons themselves. It has been suggested that the LEU–ENK terminals are derived from collaterals of parasympathetic neurons that innervate the bladder detrusor (de Groat and Booth 1984), and, perhaps, from similar neurons that innervate the smooth muscle of the colon through connections with the enteric nervous system established from the parasympathetic innervation of the myenteric plexus (Auerbach's plexus). The latter probably contain both LEU–ENK and SST. These peptides may be important in mediating reciprocal inhibition of urethral and anorectal sphincter neurons during defaecation and micturition. The origin of the VIP-ergic terminals is uncertain, but these may represent interneuronal connections (see Roppolo et al. 1985 for discussion).

Striated Sphincter Musculature

The puborectalis and pubo-anal sling muscles are derived, embryologically, not from the sphincter cloacae complex, but from the pelvi-caudal muscles that form the levator ani diaphragm of the pelvis. The relative predominance of Type 1 muscle fibres and the size relationships of the two muscle fibre types in the puborectalis and external anal sphincter muscles differ (Parks et al. 1977; Beersiek et al. 1979). In both these muscles and in the external urinary sphincter Type 1 fibres predominate. The innervation of the external anal sphincter and puborectalis muscles also differs. In accordance with these embryological principles the puborectalis muscle is innervated by direct S–2 and S–3 motor branches from the pelvic plexus, and the external anal sphincter by the inferior rectal branches of the pudendal nerves (Percy et al. 1981; Snooks and Swash 1986). The derivation of the motoneurons innervating the puborectalis is unknown, but may be from the motoneuronal pool adjacent to the Onuf nucleus, rather from this nucleus itself.

In addition, there is sexual dimorphism in the fibre diameters of the Type 1 and Type 2 muscle fibres in the human levator ani muscle. In women the Type 1 fibres are larger than the Type 2 fibres, the only skeletal muscle in humans in which this is the case; Type 2 fibres are larger than Type 1 fibres in all other skeletal muscles. In men, however, the normal excess of Type 2 fibre diameter is found in this muscle. It is not known whether this characteristic of the female levator ani is acquired at menarche, or whether it changes at the menopause, but it appears to be a hormone-dependent phenomenon (Beersiek et al. 1979).

Despite these embryological differences (Wood 1985), there are close functional relationships between the external anal sphincter and puborectalis muscles since both form part of the anorectal sphincter mechanism. For example, both show continuous tonic activity in the resting state, and both relax and become electrically silent during defaecation. In some patients with a peculiarly intractable syndrome of overwhelmingly severe constipation the pelvic floor musculature seems unable to relax normally, causing physiological obstruction to the outlet of the pelvic floor diaphragm. This syndrome has been termed "anismus" (Preston and Lennard-Jones 1985); the physiological basis for the failure of relaxation of the pelvic floor muscles during attempted defaecation in these patients is unknown. There is no evidence of any other defect of motor control in these patients, and there are no features of upper motor neuron lesion.

Suprasegmental Organisation of Sphincter Control Systems

The sacral nuclei subserving sphincter and detrusor activity are modulated by descending pathways from the central nervous system that traverse the spinal cord from the brain (Blaivas 1982; Hald and Bradley 1982). In the human these fibres belong to the corticospinal tract, and are situated in the most mesial portion of this tract (Nathan and Smith 1958). Ascending fibres, subserving sensation in the bladder and urethra, and presumably also in the anorectum, probably travel in the superficial ventral part of the lateral funiculus (Nathan and Smith 1951). The different paths followed by parasympathetic and sympathetic afferents, and by somatic afferents, from the bladder and anorectum are not yet fully understood.

Many attempts to locate centres for the control of micturition and defaecation in the central nervous system have been made during the last hundred years, the several supposed centres for these functions have been described in the cortex, in the basal ganglia and in the region of the third ventricle. In addition, several complex systems of neuronal circuits have been suggested as important in micturition, involving brain stem relays with so-called spinal centres. Loops of excitation and inhibition have been proposed, with relays extending through the length of the central nervous system from cortex to pons and to the sacral nuclei (for review see Hald and Bradley 1982). While structures capable of exerting the effects described in these accounts exist, there is little anatomical evidence in support of these concepts, which are themselves dependent on an underlying notion of nuclei and interconnecting telephone-like systems of relays, with fixed responses resulting from activity in any part of the system.

Excitatory effects in the basal ganglia may produce detrusor hyper-reflexia, but the urinary hesitancy and incontinence with constipation that often occurs in idiopathic Parkinson's disease is probably the result of associated degeneration in dopaminergic autonomic pathways, rather than from the lesion in the substantia nigra that characterises the disease. Connections probably exist from the anorectum and bladder to nuclei in the limbic system, hypothalamus and cerebellum.

More is known about the brain stem systems concerned with micturition (Barrington 1925; Blaivas 1982; Hald and Bradley 1982; Halstege and Kuypers

1982), and it is presumed that these pathways are closely integrated with those concerned with defaecation. A laterodorsal tegmental nucleus in the pons, rostral to the nucleus of the locus coeruleus, is probably the site of origin of the neuronal system that descends in the intermediolateral tract of the spinal cord to supply parasympathetic innervation to the sacral outflow and thus to the detrusor muscle of the bladder. This autonomic outflow connects with the neurons of Auerbach's myenteric plexus that modulates the enteric innervation of the colonic and anorectal smooth muscle. It has also been suggested that this autonomic innervation to the bladder detrusor muscle travels in the reticulospinal tract.

Bradley (see Hald and Bradley 1982) has tried to integrate current information on these neuronal systems in the control of bladder function and has suggested that there are four loop control systems, two for the detrusor muscle and two for the periurethral striated sphincter muscle; these all involve connections through the pudendal nuclei and the pointine brain stem. Loop 1 connects the brain stem to the frontal lobe, loop 2 connects detrusor muscle afferents to the brain stem, loop 3 consists of detrusor afferents connecting with the pudendal nucleus in the sacral cord, and loop 4 consists of afferents and efferents connecting the periurethral striated muscles with the pudendal and pontine nuclei. Clearly, this suggestion is not entirely supported by the known neurophysiology and neuroanatomy, particularly since it excludes any understanding of the interconnections of the smooth and striated innervations of the sphincter mechanisms, and does not explain the complex coordinated mechanisms required for successful urinary and faecal continence, and for micturition and defaecation.

Barrington (1921, 1931) described brain stem reflexes induced by bladder distention, and by running water through or distending the urethra. In addition, he recognised that thoracic spinal cord section would temporarily abolish micturition and that low thoracic section would permanently abolish it. These features are observed in spinal man, and the pathways suggested by Barrington for these reflexes have been largely confirmed (Denny-Brown and Robertson 1933, 1935) most recently by evoked potential and morphological studies.

The cortical localisation of the functions of micturation and defaecation was placed by Foerster, following work of Kleist (1922) on head injuries sustained by German soldiers in the First World War, on the medial surface of the cerebral hemisphere in the paracentral lobule, just anterior to the central sulcus. This localisation is consistent with that of the sensory representation of the sacral dermatomes on the cortical surface. It has recently been confirmed in normal subjects in experiments using transcutaneous electrical stimulation of the motor cortex (Merton 1985). The cortical localisation of the sensory input from these organs is less certain but is probably in the adjacent sensory cortex, also on the medial surface of the hemisphere. The localisation of the sensory input from the pelvic area can be investigated using cerebral evoked potential studies, evoked from stimulation of the pudendal nerve (dorsal nerve of the penis or clitoris), or of the pelvic detrusor nerve. Stimulation of the latter nerve results in a response in the frontal lobe, but the cortical localisation of the parasympathetic detrusor innervation of the bladder and bowel is not known. The implications of these localisations are even less certain in clinical practice since lesions of the superior parts of the medial surface of the hemispheres in the Rolandic areas are uncommon. However, frontal lobe lesions produce characteristic effects on bladder and bowel continence. These clinical syndromes overshadow the effects of lesions in other parts of the brain, since they are so common, occurring especially frequently as a result of stroke, subarachnoid haemorrhage from anterior communicating aneurysm, frontal lobe tumours and trauma (see Andrew and Nathan 1964; Maurice-Williams 1974). Kuroiwa et al. (1987) found that lesions of the right hemisphere were more likely to cause urgency and frequency of micturition than lesions of the left hemisphere. Lesions in other parts of the brain are less likely to cause major problems with continence, and only lesions in the spinal cord or conus medullaris are likely neurological causes of urinary or faecal retention.

A major problem in urology is understanding detrusor/sphincter dyssynergia. This is a functional disorder of micturition in which there is co-contraction of the bladder detrusor muscle and the urethral sphincter muscles, including the striated component. This results in functional obstruction to micturition with high bladder pressures, and to the development of potentially serious dilatation of the upper urinary tract and even to hydronephrosis. The cause of this syndrome is unknown, but it may be a feature of spinal cord lesions, for example in multiple sclerosis or spinal cord injury. It is uncommon in cerebral lesions. However, in the majority of patients full investigation fails to reveal a neurological cause. It is tempting to consider that this syndrome has its counterpart in the anorectum in the syndrome of functionally obstructed defaecation, anismus (Preston and Lennard-Jones 1985).

Investigation of Neural Pathways Subserving Pelvic Sphincters in the Human

The corticospinal fibres of the spinal cord can be excited by transcutaneous electrical stimulation, using the method introduced by Merton and Morton (1980). Stimulation can be achieved at the lumbar (T12/L1 and L4) and mid-thoracic and cervical (C6) levels, thus allowing calculation of the motor conduction velocity in between these levels. Using these methods, and recording from the puborectalis or external anal sphincter muscles, we found that the mean corticospinal spinal motor conduction velocity in 21 normal subjects, aged 22–75 years (mean 55 years) between the C6 and L1 vertebral levels was 67.4 m/s (SD 9.1 m/s) (Snooks and Swash 1985). In these subjects the mean motor conduction velocity in the cauda equina nerve roots between the T12/L1 and L4 vertebral levels, to the same muscle (puborectalis), was 57.9 m/s (SD 10.3 m/s), a difference of about 10 m/s (Kiff and Swash 1984; Snooks and Swash 1985; Swash and Snooks 1986). Thus the motor conduction velocity in the cauda equina nerve roots is similar to that in motor nerves in other parts of the peripheral nervous system. Although there was a trend for the spinal motor conduction velocity to decrease with increasing age, this was not statistically significant.

Since measurement of the length of the spinal cord may be significantly overestimated by surface measurements (Desmedt and Cheron 1983), a correction factor of about 13% may be applied in order to obtain a more accurate measure of spinal motor conduction velocity. This gives a mean motor conduction velocity in the spinal cord in normal human subjects of 58.6 m/s (Snooks and Swash 1985).

Fastest motor conduction velocity in the spinal cord represents transmission of a descending volley in the corticospinal tracts. Other descending pathways, such as the reticulospinal and rubrospinal tracts are also probably activated by the stimulus but are not likely to produce a short-latency, evoked action potential in the striated musculature. Although it is conceivable that antidromic conduction in the dorsal columns might reflexly activate lumbosacral spinal motor neurons, this is unlikely since the latency measurements do not allow for the interposition of another synaptic delay in the response mediated by C6 stimulation. Most of the nerve fibres in the human corticospinal tract are myelinated (DeMyers 1959), and their diameters range from 1 to 22 μm, but only about 2% of these fibres, corresponding to the fibres originating from the Betz cells in cortical area 4, are greater than 11 μm in diameter Lassek 1942; Brodal 1981). Applying Hursch's factor of 6 (Hursch 1939), these larger diameter fibres would be expected to conduct at velocities in the range we have recorded. In the cat, conduction velocity in the corticospinal tract ranges from 7 to 70 m/s, with peaks at 14 m/s and 42 m/s. About 3% of corticospinal fibres in the cat spinal cord conduct at velocities faster than 60 m/s (Hursch 1939).

Demyelination results in slowed conduction because of loss of saltatory conduction in the region of damaged nodes, or even in conduction block. It would therefore be expected that demyelination in the human corticospinal tract would cause slowing of conduction velocity. This could be shown using sphincter muscle recordings after spinal (C6) and conus (T12/L1) stimulation in patients with multiple sclerosis (Snooks and Swash 1985). Spinal motor conduction velocity in the corticospinal pathway to the sphincter musculature was slowed to a similar degree in these patients to that found in recordings made to lower limb muscles in patients with spastic paraparesis and extensor plantar responses due to multiple sclerosis (Cowans et al. 1984; Mills and Murray 1985; Ingram and Swash 1987). Similar observations were noted in patients with motor neuron disease in whom slowing of spinal motor conduction also correlated with clinical signs of corticospinal tract disease (Ingram and Swash 1988).

Cortical Stimulation and Sphincter Assessment

The anal and urinary sphincter muscles can be excited by cortical stimulation using electrical or magnetic stimulation. The cortical area subserving the representation of these sphincter muscles is located on the mesial surface of the hemisphere, as mapped in the experiments of Foerster and Penfield. This cortical representation can be excited when stimulation is applied across the vertex, thus directing or inducing current deep in the interhemispheric cleft. Voluntary activation of these muscles by the cooperative subject above the normal background level of excitation characteristic of these muscles serves to enhance the amplitude, and more accurately define the fastest latency of the response; the latter is usually found at about 21 ms (see Merton 1985).

If the perineal and pudendal nerve terminal motor latencies (Snooks and Swash 1984a, b; Kiff and Swash 1984) and the latency from L1 (Kiff and Swash 1984; Snooks et al. 1984, 1985a–c; Swash et al. 1985) are also measured, the latencies for central conduction in the motor pathway from the motor cortex to the C6 level and in the spinal cord to the conus medullaris at L1 level, and then in the peripheral component of the motor pathway to the sphincter muscles can be measured, and the site of any lesion accurately localised. These techniques are thus powerful in the investigation of sphincter disorders (Swash et al. 1985; Swash et al. 1987), as well as in the investigation and management of other neurological disorders.

Acknowledgements

I would like to thank the research fellows with whom I have been privileged to work, who have come from many different countries to work in the Sir Alan Parks Laboratory at St Mark's Hospital to address the problem of sphincter disorders: P. Barnes, F. Beersiek, D. Chalmers, G. Browning, M. M. Henry, P. Jones, E. S. Kiff, S. Laurberg, D. Z. Lubowski, S. Mathers, M. E. Neill, J. Percy, M. Pescatori, S. J. Snooks, T. Teramoto and M. Wunderlich. The work has been supported throughout by the St Mark's Hospital Research Fund, and by the Sir Alan Parks Research Fund of the Royal College of Surgeons of England, together with many other generous benefactions.

References

Andrew J. Nathan P (1964) Lesions of the anterior frontal lobes and disturbances of micturition and defaecation. Brain 87:233–262

Appenzeller O, Atkinson R (1985) The autonomic nervous system. In: Swash M, Kennard C (eds) Scientific basis of clinical neurology. Churchill Livingstone, London, pp 463–488

Barrington FJF (1921) The relation of the hind-brain to micturition. Brain 44:23–53

Barrington FJF (1925) The effect of lesion of the hind and midbrain on micturition in the cat. Q J Exp Physiol 15:81–102

Barrington FJF (1931) The component reflexes of micturition in the cat. Brain 54:177–188

Bartram CI, Mahieu PHG (1985) Radiology of the pelvic floor. In: Henry MM, Swash M (eds) Coloproctology and the pelvic floor. Butterworths, London, pp 151–186

Beersiek F, Parks AG, Swash M (1979) Pathogenesis of idopathic anorectal incontinence; a histometric study of the anal sphincter musculature. J Neurol Sci 42:111–127

Blaivas JG (1982) The neurophysiology of micturition; a clinical study of 550 patients. J Urol 127:948–963

Brodal A (1981) Neurological anatomy in relation to clinical medicine. Oxford University Press, Oxford

Cowans JMA, Dick JPR, Day BL, Rothwell JC, Thompson PD, Marsden CD (1984) Abnormalities in central motor pathway conduction in multiple sclerosis. Lancet II:304–307

de Groat WC, Booth AM (1984) Autonomic systems to the urinary bladder and sexual organs. In: Dyck PJ, Thomas PK, Lambert EH, Bunge R (eds) Peripheral neuropathy. Saunders, Philadelphia, pp 285–299

DeMyer W (1959) Number of axons and myelin sheaths in the adult human medullary pyramids; study with silver impregnation and iron haematoxylin staining methods. Neurology (Minneap) 9:42–47

Denny-Brown D, Robertson EG (1933) On the physiology of micturition. Brain 56:149–190

Denny-Brown D, Robertson EG (1935) An investigation of the nervous control of defaecation. Brain 98:256–310

Desmedt JE, Cheron G (1983) Spinal and far-field components of human somatosensory evoked potentials to posterior tibial nerve stimulation analysed with oesphageal derivations and non-cephalic reference recording. Electroencelphalogr Clin Neurophysiol 56:635–651

Floyd WF, Walls EW (1953) Electromyography of the sphincter ani externus in man. J Physiol 122:599–609
Gillis R, Norman W, Kasbekar D, Mangel A, Skirboll L, Quest J, Payani F (1987) Central nervous system control of gastrointestinal function. In: Szurszewski JH (ed) Cellular physiology and clinical studies of gastrointestinal function. ICS 725 Excerpta Medica, Amsterdam, pp 209–226
Gosling J (1979) The structure of the bladder and urethra in relation to function. Urol Clin North Am 6:31–38
Gowers WR (1877) The automatic action of the sphincter ani. Proc R Soc Med 26:77
Hald T, Bradley WE (1982) The urinary bladder. Williams and Wilkins, Baltimore, p 339
Henry MM, Swash M (1985) Coloproctology and the pelvic floor. Butterworths, London
Halstege G, Kuypers HGJM (1982) The anatomy of the brain stem pathways to the spinal cord in cat; a labelled amino acid tracing study. In: Kuypers HGJM, Martin GF (eds) Descending pathways to the spinal cord. Elsevier, Amsterdam, pp 146–175 (Progress in brain research, vol 57)
Hursch JB (1939) Conduction velocity and diameter of nerve fibres. Am J Physiol 127:131–139
Ingram DA, Swash M (1987) Central motor conduction is abnormal in motor neuron disease. J Neurol Neurosurg Psychiatry 50:159–166
Ingram DA, Swash M (1988) Central motor conduction in multiple sclerosis; evaluation abnormalities revealed by electromagnetic stimulation of the brain. J Neurol Neurosurg Psychiatry 51:487–494
Ito M (1986) Neural systems controlling movement. Trends Neurosci 9:515–518
Kiff ES, Swash M (1984) Normal proximal and delayed distal conduction in the pudendal nerves of patients with idiopathic (anorectal) incontinence. J Neurol Neurosurg Psychiatry 47:820–823
Kleist K (1922) Gehirnpathologie. In: Handbuch der artzlichen Erfahrungen im Weltkrieg 1914–1918, Vol 1V. Barth. Leipzig
Kuroiwa Y, Tohgi H, Ono S, Itoh M (1987) Frequency and urgency of micturition in hemiplegic subjects; relationship to hemisphere laterality of lesions. J Neurol 234:100–102
Lassek AM (1942) The human pyramidal tract. IV. A study of the mature, myelinated fibres of the pyramid. J Comp Neurol 76:217–225
Lubowski DZ, Nicholls RJ, Swash M, Jordan MJ (1987) Neural control of internal anal sphincter function. Br J Surg 74:668–670
Maurice-Williams RS (1974) Micturition symptoms in frontal tumours. J Neurol Neurosurg Psychiatry 37:431–436
McKenna KE, Nadelhaft I (1984) Organization of the pudendal nerves in the male and female rat. Soc Neurosci (abstr) 10:902
Merton PA (1985) Electrical stimulation through the scalp of pyramidal fibres supplying pelvic floor muscles. In: Henry MM, Swash M (eds) Coloproctology and the pelvic floor. Butterworths, London, pp 125–128
Merton PA, Morton HB (1980) Stimulation of the cerebral cortex in the intact human subject. Nature 285: 227
Mills KR, Murray NMF (1985) Corticospinal tract conduction time in multiple sclerosis. Ann Neurol 18:601–605
Nathan P, Smith M (1951) The centripetal pathway from the bladder and urethra within the spinal cord. J Neurol Neurosurg Psychiatry 14:262–282
Nathan P, Smith M (1958) The centrifugal pathway for micturition in the spinal cord. J Neurol Neurosurg Psychiatry 21:177–189
Onuf (Onufrowicz) B (1900) On the arrangement and function of the cell groups in the sacral region of the spinal cord. Arch Neurol Psychopathol 3:387–411
Parks AG (1975) Anorectal incontinence. Proc R Soc Med 68:681–690
Parks AG, Porter NH, Melzack J (1962) Experimental study of the reflex mechanisms controlling the muscles of the pelvic floor. Dis Colon Rectum 5:407–414
Parks AG, Swash M, Urich H (1977) Sphincter denervation in anorectal incontinence and rectal prolapse. Gut 18:656–665
Percy JP, Neill ME, Swash M, Parks AG (1981) Electrophysiological study of motor nerve supply of pelvic floor. Lancet I:16–17
Preston DM, Lennard-Jones JE (1985) Anismus in chronic constipation. Digest Dis Sci 30:413–418
Roney KJ, Scheibel AB, Shaw GL (1979) Dendritic bundles; survey of anatomical experiments and physiological theories. Brain Res Rev 1:225–271
Roppolo JR, Nadelhaft I, de Groat WC (1985) The organization of pudendal motoneurons and primary afferent projections in the spinal cord of the rhesus monkey revealed by horseradish peroxidase. J Comp Neurol 234:475–488

Schroder HD (1980) Organisation of the motor neurones innervating the pelvic muscles of the male rat. J Comp Neurol 192:567–587
Schroder HD (1984) Somatostatin in the caudal spinal cord; an immunohistochemical study of the spinal centres involved in the innervation of pelvic organs. J Comp Neurol 223:400–414
Snooks SJ, Swash M (1984a) Abormalities of the innervation of the urethral striated sphincter musculature in incontinence. Br J Urol 56:401–405
Snooks SJ, Swash M (1984b) Perineal nerve and transcutaneous spinal stimulation; new methods for the investigation of the urethral striated sphincter musculature. Br J Urol 56:406–409
Snooks SJ, Swash M (1985) Motor conduction velocity in the human spinal cord; slowed conduction in multiple sclerosis and radiation myelopathy. J Neurol Neurosurg Psychiatry 48:1135–1139
Snooks SJ, Swash M (1986) The innervation of the muscles of continence. Ann R Coll Surg Engl 68:45–49
Snooks SJ, Barnes PRH, Swash M (1984) Damage to the innervation of the voluntary anal and periurethral striated sphincter musculature in incontinence; an electrophysiological study. J Neurol Neurosurg Psychiatry 47:1269–1273
Snooks SJ, Badenoch D, Tiptaft R, Swash M (1985a) Perineal nerve damage in genuine stress urinary incontinence. Br J Urol 57:422–426
Snooks SJ, Barnes PRH, Swash M, Henry MM (1985b) Damage to the innervation of the pelvic floor musculature in chronic constipation. Gastroenterology 89:971–981
Snooks SJ, Swash M, Henry MM (1985c) Abnormalities in peripheral and central nerve conduction in anorectal incontinence. J R Soc Med 78:294–300
Swash M (1985) Histopathology of the pelvic floor. In: Henry MM, Swash M (eds) Coloproctology and the pelvic floor. Butterworths, London, pp 129–150
Swash M, Snooks SJ (1985) Electromyography in pelvic floor disorders. In: Henry MM, Swash M Coloproctology and the pelvic floor. Butterworths, London, pp 85–103
Swash M, Snooks SJ (1986) Slowed motor conduction in lumbosacral nerve roots in cauda equina lesions: a new diagnostic technique. J Neurol Neurosurg Psychiatry 49:808–816
Swash M, Snooks SJ, Henry MM (1985) A unifying concept of pelvic floor disorders and incontinence. J R Soc Med 78:906–911
Swash M, Snooks SJ, Chalmers DHK (1987) Parity as a factor in incontinence in multiple sclerosis. Arch Neurol 44:504–508
Taverner D, Smiddy J (1959) An electromyographic study of the normal function of the external anal sphincter and pelvic diaphragm. Dis Colon Rectum 2:153–160
Winckler G (1958) Remarques sur la morphologie et l'innervation du M releveur de l'anus. Arch Anat Histol Embryol (Strasb) 41:77–95
Womack NR, Williams NS, Holmfield JHM, Morrison JFB, Simpkins KC (1987) New method for the dynamic assessment of anorectal function in constipation. Br J Surg 85:994–998
Wood B (1985) Anatomy of the anal sphincters and pelvic floor. In: Henry MM, Swash M (eds) Coloproctology and the pelvic floor. Butterworths, London, pp 3–21

Subject Index